Encyclopedia of Herbicides: Methods and Mode of Action

Volume I

Encyclopedia of Herbicides: Methods and Mode of Action Volume I

Edited by **Molly Ismay**

New York

Published by Callisto Reference,
106 Park Avenue, Suite 200,
New York, NY 10016, USA
www.callistoreference.com

Encyclopedia of Herbicides: Methods and Mode of Action
Volume I
Edited by Molly Ismay

International Standard Book Number: 978-1-63239-255-8 (Hardback)

Contents

Preface

Herbicide use is an essential part of agricultural practices for optimizing crop production these days. This book broadly covers the mechanisms of herbicidal action, mode of action of certain herbicides on controlling diseases, weed expansion and production, and growth and progress of field crops. This compilation is well elaborated and summarized with topics like molecular mechanism of action, immunosensors, community response, utilization of herbicides in biotech culture, weed resistance, herbicides risk and herbicides persistence. The studies in this book reflect that practice with one crop or problem at times can be suitable to a seemingly different situation in a different crop often till an unexpected extent. The topics covered in this book will appeal to readers concerned with fields related to herbicides and pesticides and the content will be of great interest to them in their future prospects.

Significant researches are present in this book. Intensive efforts have been employed by authors to make this book an outstanding discourse. This book contains the enlightening chapters which have been written on the basis of significant researches done by the experts.

Finally, I would also like to thank all the members involved in this book for being a team and meeting all the deadlines for the submission of their respective works. I would also like to thank my friends and family for being supportive in my efforts.

Editor

Part 1

Physiological and Molecular Mechanisms

Molecular Mechanism of Action of Herbicides

Istvan Jablonkai

Institute of Biomolecular Chemistry, Chemical Research Center,
Hungarian Academy of Sciences, Budapest,
Hungary

1. Introduction

Herbicides are the most widely used class of pesticides accounting for more than 60% of all pesticides applied in the agriculture (Zimdahl, 2002). The herbicide's mode of action is a biochemical and physiological mechanism by which herbicides regulate plant growth at tissue and cellular level. Herbicides with the same mode of action generally exhibit the same translocation pattern and produce similar injury symptoms. At the physiological level, the various herbicides control plants by inhibiting photosynthesis, mimicking plant growth regulators, blocking amino acid synthesis, inhibiting cell elongation and cell division, etc.

There are approximately 20 different target sites for herbicides (Shaner, 2003). Despite the relative importance of herbicides within crop protection products only a low number of biochemical mode of action can be shown for the marketed herbicides. Herbicides with 6 mode of action represent around 75% of herbicide sales (Klausener et al., 2007). Understanding the mode of action of herbicides has been an important tool in research to improve application methods in various agricultural practices, handle weed resistance problems and explore toxicological properties. Several enzymes and functionally important proteins are targets in these biochemical processes. Classical photosystem-II (PSII) inhibitors bind to D1 protein, a quinone-binding protein to prevent photosynthetic electron transfer. Inhibition of biosynthesis of aromatic amino acids relies on the enzyme 5-enolpyruvylshikimate 3-phosphate (EPSP) synthase. Acetohydroxyacid sythase (AHAS), a target of several classes of herbicides catalyzes the first common step in the biosynthesis of valine, leucine, and isoleucine. Several different types of herbicides apparently cause accumulation of photodynamic porphyrins by inhibiting protoporphyrinogen oxidase (PPO). Formation of homogentisate via inhibition of 4-hydroxyphenylpyruvate dioxygenase (HPPD), a key enzyme in tyrosine catabolism and carotenoid synthesis inhibited by herbicides having different structure. Lipid biosynthesis is the site of action of a broad array of herbicides used in controlling monocot weeds by inhibiting acetyl-CoA carboxylase (ACC) or very-long-chain fatty acids (VLCFA). Several compounds are potent inhibitors of glutamine synthase (GS) which catalyzes the incorporation of ammonia into glutamate.

The decreasing heterogenity of herbicides targeting fewer mechanism of action is increasing the prevalence of herbicide resistance (Lein et al., 2004). Therefore, characterization of new modes of action by exploring novel targets is of high importance for discovery of new compound classes. Elucidation of the atomic structure of target site proteins in complex with herbicides is important for understanding the initial biochemical response following application. Furthermore, the knowledge of molecular mechanism of action may provide a powerful tool to manipulate herbicide selectivity and resistance. *De novo* design of potent enzyme inhibitors has increased dramatically, particularly as our knowledge of enzyme reaction mechanisms has improved. Recent findings on the interaction of herbicides with target site enzymes and receptor proteins involved in their mode of action will be reviewed in this chapter.

2. Target site action of herbicides

2.1 Interaction of amino acid biosynthesis inhibitor herbicides with target site enzymes

2.1.1 Aromatic amino acid biosynthesis inhibitors

Inhibitors of biosynthesis of aromatic amino acids such as phenylalanine, tyrosine and tryptophan target the shikimic acid pathway. The first step of the synthesis of these three amino acids is the condensation of D-erythrose 4'-phosphate with phosphoenolpyruvate (PEP) to produce 3'-deoxy-D-arabino-heptulosonic acid 7'-phosphate (Figure 1). This undergoes a series of reactions, including loss of a phosphate, ring closure and a reduction to give shikimic acid, which is then phosphorylated by shikimate kinase. Shikimate phosphate is combined with a further molecule PEP to give 3-enolpyruvylshikimate 5-phosphate (EPSP). The enzyme EPSP synthase catalyzes the transfer of the enolpyruvyl moiety of PEP to the 5-hydroxyl of shikimate-3-phosphate (S3P) (Amrhein et al., 1980) has

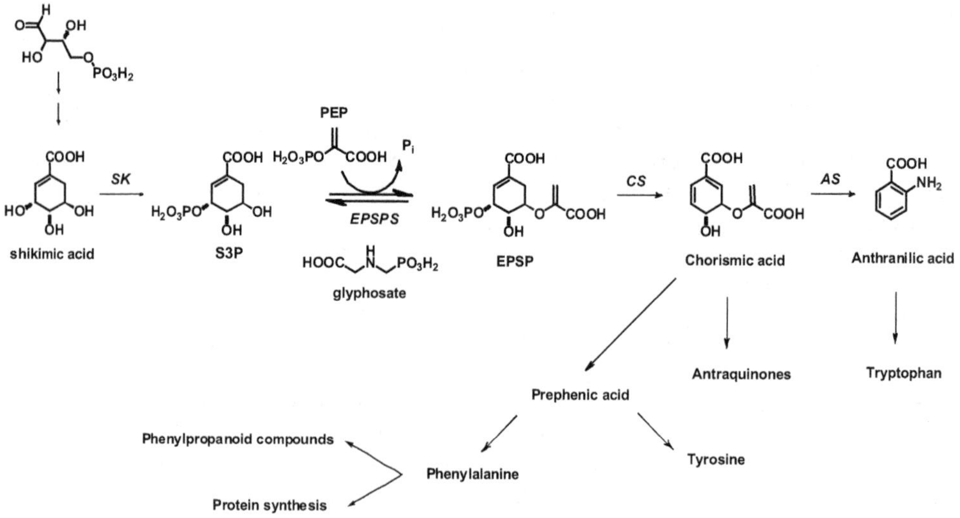

Fig. 1. Shikimic acid pathway. Biosynthesis of aromatic amino acids and action of the herbicide glyphosate. SK= shikimate kinase, EPSPS= 5-enolpyruvyl-shikimate-3-phosphate synthase, CS= chorismate synthase, AS= anthranilate synthase.

received considerable attention because it is inhibited by the herbicide, glyphosate. EPSP is converted to chorismic acid, which is at a branch point in this pathway, and can undergo two different reactions, one leading to tryptophan, and the other to phenylalanine and tyrosine. The broad-spectrum herbicide glyphosate, the active ingredient of Round-up, inhibits EPSP synthase, the enzyme catalyzing the penultimate step of the shikimate pathway toward the biosynthesis of aromatic amino acids. The extraordinary success of this simple and small molecule is based on its high specificity for plant EPSP enzymes (Pollegioni et al., 2011).

The first crystal structure of EPSPS was determined for the *E. coli* enzyme in its ligand-free state (Stallings et al., 1991). EPSP synthase (M_r 46,000) folds into two globular domains, each comprising three identical βαβαββ-folding units connected to each other by a two-stranded hinge region. The structure upon interaction of EPSP synthase from *E. coli* with one of its two substrates (S3P) and with glyphosate was identified a decade later (Schönbrunn et al., 2001). The two-domain enzyme was shown to close on ligand binding, thereby forming the active site in the interdomain cleft. Glyphosate occupied the binding site of the second substrate PEP of EPSP synthase, mimicking an intermediate state of the ternary enzyme-substrates complex. (Figure 2). The glyphosate binds close to S3P without perturbing the structure of active-site cavity. The 5-hydroxyl group of S3P was found hydrogen-bonded to the nitrogen atom of of the herbicide and the glyphosate binding site is dominated by charged residues from both domains of the enzyme, of which Lys-22 (K22), Arg-124 (R124) and Lys-411 (K411) was found in the PEP binding (Shuttleworth et al., 1999). Gly-96 (G96) residue which is not the most important in the herbicide binding plays a key role in glyphosate sensitivity of plants since replacing it an alanine residue provides the glyphosate-tolerant mutant protein (Sost and Amrhein, 1990).

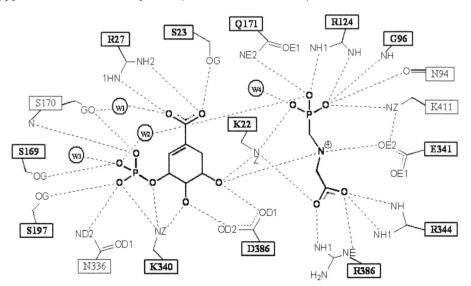

Fig. 2. Schematic representation of ligand binding in EPSP synthase-S3P-glyphosate complex (Schönbrunn et al., 2001). Ligands are drawn in bold lines. Dashed lines indicate H-bonds and ionic interactions. Strictly conserved residues are highlighted by bold labels.

Round-up ready crops such as maize, soybean, cotton and canola carry the gene coding for a glyphosate-insensitive form of EPSPS enzymes which enables more effective weed control by allowing postemergent herbicide application (Padgette et al., 1995). The genetically engineered maize lines NK603 and GA21 carry carry distint EPSPS enzymes. NK603 maize line contains a gene derived from *Agrobacterium* sp. strain CP4 encoding a glyphosate tolerant class II enzyme, the so-called CP4 EPSP synthase. On the other hand GA21 maize was created by point mutations of class I EPSPS such as enzymes from *Zea mays* and *E. coli* which are sensitive to low glyphosate concentrations. Although these crops have been widely used, the molecular basis for the glyphosate-resistance has remained obscure.

The three-dimensional structure of CP4 EPSP synthase revealed that the enyzme exists in an open, unliganded state (Funke et al., 2006). Upon interaction with S3P, the enzyme undergoes a large conformational change suggesting an induced-fit mechanism with binding of S3P as a prerequisite for the enzyme's interaction with PEP. During interaction with glyphosate the herbicide binds to the active site of CP4 EPSP adjacent to S3P. The weak action of glyphosate on CP4 EPSP synthase can be primarily attributed to an Ala residue in position 100 of which methyl group protrudes into the glyphosate binding site and clashes with one of the oxygen atoms of the herbicide phosphonate group. As a result, the glyphosate molecule adopts a substantially different shortened conformation as interacts with the CP4 enzyme (Figure 3). Replacing Ala-100 with a Gly allows glyphosate to bind in its extented conformation positioning its N atom midway between the target hydroxyl of

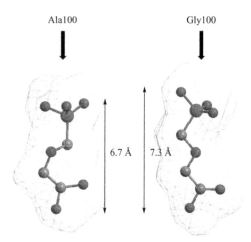

Fig. 3. Shortened and extended conformation of glyphosate (Funke et. al, 2006). Left, with Ala residue in position 100 the herbicide is ~0.6 Å shorter.

S3P and Glu-354. The mutation of Ala-100 to Gly restored the CP4 enzyme's sensitivity toward glyphosate. It appears that the conformational change introduced upon glyphosate binding simple makes the EPSPS active site unavailable to PEP. Based on this molecular basis for glyphosate resistance a novel inhibitors of EPSP synthase can be designed in case of emergence of glyphosate-resistant weeds. Nevertheless, structure-activity relationships on the inhibition of EPSP synthase with analogs of glyphosate revealed that minor structural alterations resulted in dramatically reduced potency and no compound superior to glyphosate was identified (Franz et al., 1997; Sikorski and Gruys, 1997; Mohamed Naseer Ali et al., 2005).

Molecular basis for glyphosate-tolerant GA21 maize resulting from the double mutation Thr-97→Ile and Pro-101→Ser (T97I/P101S, TIPS) and single mutation (T97I) in EPSPS from *E. coli* has recently been revealed (Funke et. al, 2009). The crystal structure of EPSPS demonstrated that the dual mutation causes a shift of residue Gly-96 toward the glyphosate binding site, impairing efficient binding of glyphosate, while the side chain of Ile-97 points away from the substrate binding site, facilitating PEP utilization. The single site T97I mutation renders the enzyme sensitive to glyphosate and causes a substantial decrease in the affinity for PEP. Thus, only the concomitant mutations of Thr-97 and Pro-101 induce the conformational changes necessary to produce catalytically efficient, glyphosate-resistant class I EPSPS. Mutations of the residue corresponding to Pro-101 of *E. coli* EPSPS have been reported in a number of field-evolved glyphosate-resistant weeds (Yu et al., 2007; Perez-Jones et al., 2007). However, mutations of Thr-97 have never been observed. The decreased catalytic efficiency of the T97I mutant EPSPS with respect to utilization of PEP may explain why it has not been observed in glyphosate resistant weeds.

Detoxication of the glyphosate by oxidases and acetyltransferase has been a promising strategy to confer resistance (Pollegioni et al., 2011). However, none of these mechanisms has been shown to occur in higher plants to a significant degree. The metabolism by glyphosate oxidoreductase (GOX) and glycine oxidase (GO) resulting in the formation of aminomethyl-phosphonic acid (AMPA) and glyoxylate (the AMPA pathway) takes place only in soil by a number of Gram-positive and Gram-negative bacteria. Chemical mutagenesis and error-prone PCR were used to insert genetic variability in the sequence coding for GOX and the enzyme variants were selected for their ability to grow at glyphosate concentrations that inhibit growth of the *E. coli* methylphosphonate-utilizing control strain (Barry and Kishore, 1998). The best variants had a more basic residue at position 334. However the low level of activity and heterologous expression observed for GOX might explain the limitations encountered in developing commercially available crops based on this enzyme. Furthermore, GO can be efficiently expressed as an active and stable recombinant protein in *E. coli* (Job et al., 2002). Because of the introduction of an arginine at position 54 the crystal structure of the multiple-point variant G51S/A54R/H244A has a different conformation from the wild-type GO. The presence of a smaller alanin at position 244 eliminates steric clashes with the side chain of Glu-55 thus facilitating the interaction between Arg-54 and glyphosate (Pedotti et al., 2009). Glyphosate acetyltransferase (GLYAT) is an acetyltransferase from *Bacillus licheniformis* that was optimized by gene shuffling for acetylation of glyphosate paving the way for the development of glyphosate tolerance in transgenic plants (Castle et al., 2004). The catalytic action of GLYAT requires a cofactor AcCoA. Four active site residues (Arg-21, Arg-73, Arg-111, and His-138) contribute to a positively charged substrate-binding site (Siehl et al., 2007). His-138 functions as a catalytic base via substrate-assisted deprotonation of the glyphosate secondary amine, whereas another active site residue Tyr-118 functions as a general acid.

Despite successful efforts on developing glyphosate-resistant crops there are increasing instances of evolved glyphosate resistance in weed species (Waltz, 2010). In order to preserve the utility of this valuable herbicide, growers must be equipped with effective and economic herbicide-trait combinations to use in rotation or in combination with glyphosate (Pollegioni et al., 2011).

2.1.2 Acetohydroxyacid synthase (AHAS) inhibitors

The endogenous AHAS gene is involved in the biosynthesis of branched chain amino acids (valine, leucine and isoleucine) catalyzing the formation of 2-acetolactate or 2-aceto-2-hydroxybutyrate (Duggleby and Pang, 2000) (Figure 4). AHAS is the site of action of several structurally diverse classes of herbicides such as sulfonylureas (La Rossa and Schloss, 1984), imidazolinones (Shaner, 1984), triazolopyrimidine sulfonamides (Gerwick et al., 1990). These herbicides are unusual inhibitors since they do not exhibit structural similarity to substrates (pyruvate, α-ketobutyrate), cofactors (thiamine diphosphate (ThDP), FAD) and allosteric effectors (valine, leucine and isoleucine) of the enzyme. When AHAS is inhibited, deficiency of the amino acids causes a decrease in protein synthesis leading to reduced cell division rate (Rost, 1984; Shaner and Singh, 1993). This process eventually kills the plants after showing symptoms in meristematic tissues where biosynthesis of amino acids primarily takes place (Zhou et al., 2007).

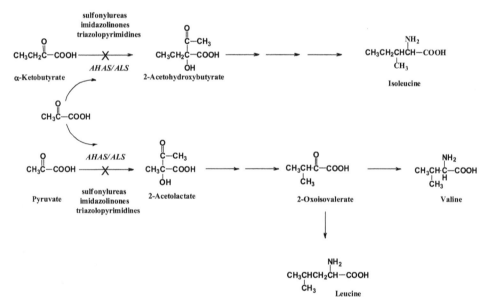

Fig. 4. Biosynthetic pathway of branched chain amino acids and the site of action of herbicidal inhibitors.

The crystalline structure of any plant protein in complex with a commercial herbicide was reported first for *Arabidopsis thaliana* AHAS in complex with the sulfonylurea herbicide chlorimuron ethyl (Pang et al., 2004). There was one monomer in the asymmetric unit and these were arranged as pairs of dimers in the crystal. The dimers form a very open hexagonal lattice, with a high solvent content of 81 %. The 3D structure of *Arabidopsis thaliana* AHAS in complex with five sulfonylureas and with the imidazolinone, imazaquin has been published later by the same research group (McCourt et al., 2006). The *At*AHAS is a tetramer consisting of four identical subunits with an overall fold. Each subunit has three domains and a C-terminal tail that loops over the active site. Associated with each subunit is FAD, Mg-ThDP, >200 water molecules and one molecule of sulfonylurea or two of imazaquin. A prolyl *cis* peptide bond observed between Leu-648 and Pro-649 at the C-terminal tail. Pro-649 is completely conserved in AHAS from 21 species (Duggleby and Pang, 2000) suggesting the critical function of this residue when the C-terminal tail changes from a disordered state in its free structure to the ordered state during the catalytic cycle. Neither sulfonylureas nor imazaquin have a structure that mimics the substrates for the enzyme, but both inhibit by blocking a channel through which access to the active site is gained. In binding of sulfonylureas to plant AHAS a bend at the sulfonyl group positions the two rings almost orthogonal to each other. The sulfonyl group and the adjacent aromatic ring are situated at the entrance to a channel leading to the active site with the rest of the molecule inserting into the channel. In *At*AHAS-imazaquin complex two herbicide molecules was found to bind to each subunit. One of these is within the channel leading to the active site, whereas a second is located around 20 Å from the active site in a pocket. Ten of the amino acid residues that bind the sulfonylureas also bind imazaquin. Six additional residues interact only with the sulfonylureas, whereas there are two residues that bind imazaquin but not the sulfonylureas. Thus, the two classes of inhibitor occupy partially overlapping sites but adopt different modes of binding. The positions of several key residues (Arg-199, Asp-376, Arg-377, Trp-574, Met-200) at the entrance of active-site channel move to accomodate the sulfonylurea chlorimuron-ethyl or imazaquin (Figure 5). Overall 28 van der Waals interaction and only one hydrogen bond contribute to the binding of imazaquin while 50 van der Walls contacts and six hydrogen bonds make a stronger binding for chlorimuron-ethyl. The higher affinity and depeer binding of binding into the active site makes chlorimuron-ethyl more potent inhibitor ($K_{i(app)}$= 10.8 nM) to *At*AHAS as compared to imazaquin ($K_{i(app)}$= 3.0 μM).

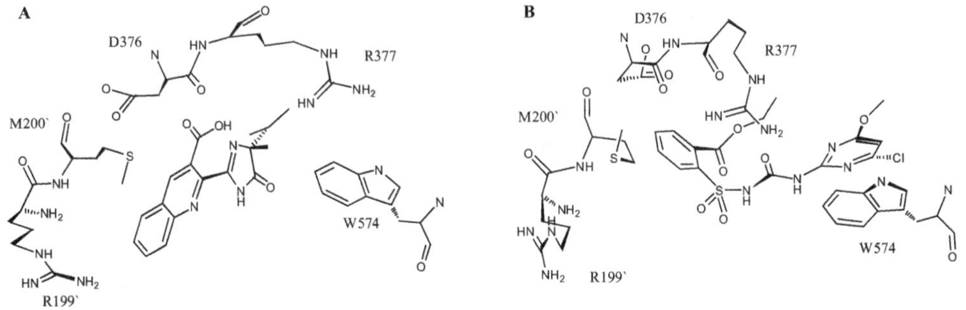

Fig. 5. Schematic representation of conformational adjustments in the *At*AHAS herbicide binding sites (McCourt et al., 2006). (A) Imazaquin. (B) Chlorimuron-ethyl.

The increasing emergence of resistant weeds due to the appearance of mutations that interfere with the inhibition of AHAS is now a worldwide problem. Knowledge of atomic resolution of the enzyme allows us to explain how the substitution of key amino acid residues by mutation results in resistantance to these herbicides. Most AHAS isoenzymes resistant to the herbicides carry substitutions for the amino acid residues Ala-122 (amino acid numbering refers to the sequence in *Arabidopsis thaliana*), Pro-197, Ala-205 located at the

N-terminal end of the enzyme whereas Asp-376, Trp-574, and Ser-653 are located at the C-terminal end (Tranel and Wright, 2002). Ala-205→Val mutation resulted in resistance in eastern black nightshade (*Solanum Ptychanthum*) (Ashigh and Tardif, 2007). Eight different amino acid substitutions of Pro-197 have been found to confer herbicide resistance but only Pro-197→Leu has been implicated in strong resistance to imidazolinones (Sibony et al., 2001). It is likely that the bulky Leu residue prevents the entry of imidazolinones into the channel whereas any substitution inhibits sulfonylurea access. Ala-122→Thr (Bernasconi et al., 1995) and Ser-653→Asn (Hattori et al., 1992; Lee et al., 2011) confers strong resistance to the imidazolinones but not to sulfonylureas. Replacement of these residues by a larger one seems to impair imidazolinone binding because the steric hindrance change space where the aromatic ring situated. Substitution of Trp-574, a residue important for defining the shape of the active-site channel, by leucine changes the shape of the binding-site channel and endow high level of resistance to both both imidazolinones and sulfonylureas (Bernasconi et al., 1995).

In a recently published paper (Le et al., 2005) the role of three well-conserved arginine residues (Arg-141, Arg-372, and Arg-376) of tobacco AHAS was determined by site-directed mutagenesis. Arg-372 and 376 residues are important for catalytic activity as they affect the binding with the cofactor FAD. The mutated enzymes such as Arg-141→Ala, Arg-141→Phe and Arg-376→Phe were inactive and unable to bind the cofactor, FAD. The inactive mutants had the same secondary structure as that of the wild type. The mutants Arg-141→Lys, Arg-372→Phe, and Arg-376→Phe exhibited much lower specific activities than the wild type and moderate resistance to herbicides such as bensulfuron methyl and AC 263222. The mutation showed a strong reductions in activation efficiency by thiamine diphosphate, while mutations Arg-372→Lys and Arg-376→Lys showed a strong reduction in activation efficiency by FAD in comparison to the wild type enzyme. Results suggested that the residue Arg-141 is located at the active site and may affect the binding with cofactors while Arg-372 and Arg-376 are located at the overlapping region of the FAD-binding site and are a common binding site for the three classes of herbicides. The molecular basis for inhibition of AHAS enzymes enables us to explain evolved weed resistance and thus allowing more sophisticated AHAS inhibitors to be developed.

2.1.3 Glutamine synthetase (GS) inhibitors

GS is one of the essential enzymes for plant autotrophy catalyzes the the incorporation of the ammonia into glutamate to generate glutamine with concomitant hydrolysis of ATP. Phosphinothricin (PPT) is a potent GS inhibitor (Lydon and Duke, 1999). Actually, PPT, a metabolite of a herbicidally inactive natural product bialaphos has been registered in many countries as a non-selective herbicide. GS plays a crucial role in the assimilation and re-assimilation of ammonia derived from a wide variety of metabolic processes during plant growth and development. The first crystal structure of maize (*Zea mays* L.) GS has recently been reported (Unno et al., 2006). The structure reveals a unique decameric structure that differs significantly from the bacterial GS structure. The GS decamer contains 10 active sites and each active site is located between two adjacent subunits in a pentamer. The active sites (20 Å deep) are formed between two neighboring monomers. The phosphorylated PPT (P-PPT) binding sites were found at the bottoms of the 10 clefts. The ADP binding sites in the ADP/P-PTP/Mn complex structures and the adenylimido-diphosphate (AMPPNP) binding sites in the AMPPNP/PPT/Mn complex structure are located near the openings in the 10 catalytic clefts. The P-PPTmolecule is bound mainly by the main chain of Gly-245 and the side chains of Glu-131, Glu-192, His-249, Arg-291, Arg-311, and Arg-332 through hydrogen bond interactions in addition to three Mn^{2+} ions. The phosphate group of the P-PPT coordinates to the three Mn^{2+}. The structures of complexes revealed the mechanism for the transfer of phosphate from ATP to glutamate and to interpret the inhibitory action of phosphinothricin as a guide for the development of new potential herbicides.

2.2 Interaction of herbicides with 4-hydroxyphenylpyruvate dioxygenase (HPPD)

4-Hydroxyphenylpyruvate dioxygenase (HPPD) converts 4-hydroxyphenyl-pyruvate (HPP) into homogentisate (HGA) with the concomitant release of CO_2 is a target of β-triketone and isoxazole herbicides (Shaner, 2003). This nonheme, Fe^{2+}-containing, α-keto acid-dependent enzyme catalyzes a complex reaction involving the oxidative decarboxylation of the 2-oxoacid side-chain of 4-hydroxyphenyl-pyruvate, the subsequent hydroxylation of the aromatic ring, and a 1,2-rearrangement of the carboxymethyl group to yield homogentisic acid (Pascal et al., 1985) (Figure 6). The mechanism of this complex reaction has recently been revealed that the native HPPD hydroxylation reaction results in the formation of ring epoxide as the first intermediate (Shah et al., 2011). Homogentisic acid is a precursor in the biosynthesis of the plastoquinones and alpha-tocopherol. Plastoquinones are vital cofactors for phytoene desaturase (PDS) and their loss results in the inhibition of PDS and a decrease in carotenoid levels. The inability to offload electrons from the photosystems results in bleaching of the affected plants due to reduced chlorophyll levels. Triketone inhibitors exhibit structural similarity to the substrate HPP and therefore will bind bidentate to the active ferrous form of the enyzme (Prisbylla et al., 1993).

The first X-ray crystal structure of HPPD published was from *Pseudomonas fluorescens* (Serre et al., 1999) followed by structures from *Arabidopsis thaliana* (Yang et al., 2004; Fritze et al., 2004) *Zea mays* (Fritze et al., 2004), *Streptomyces avertilis* (Brownlee et al., 2004), and rat (Yang et al., 2004). However, the crystal structure of an HPPD from *Pseudomonas fluorescens* showed relatively low overall sequence homology to plant and mammalian HPPDs (21% and 29% amino acid identity, respectively) (Serre et al., 1999). The protein has a subunit mass of 40-50 kDa and typically associated to form dimers in eukaryotes (Moran 2005).

In HPPD structures the N- and C-termini fold into discrete domains and the active site is formed exclusively from the residues of the C-termini (Moran 2005). The peptide fold of HPPDs have a jellyroll fold motif (eight β-strands arranged in a barrel).

Crystal structures of *Arabidopsis thaliana*, *Zea mays* (Fritze et al., 2004) revealed that the C-terminal helix gates substrate access to the active site around a non-heme Fe^{2+}-containing center. In the Z. *mays* HPPD structure this helix packs into the active site, sequestering completely it from the solvent while in the *Arabidopsis* structure tilted by about 60° into the solvent and leaves the active site fully accessible. The crystal structures of the herbicidal target enzyme HPPD from the *Arabidopsis* with and without an herbicidal benzoylpyrazole inhibitor that potently inhibits both plant and mammalian HPPDs have been determined (Figure 7) (Yang et al., 2004). The active site of *At*HPPD is located within an open twisted barrel-like β sheet. In common with other members of this dioxygenase family, the required

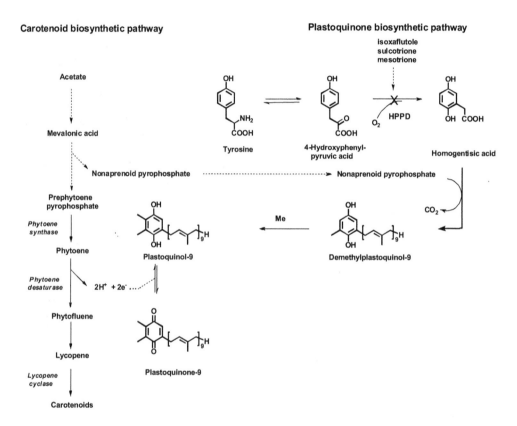

Fig. 6. Carotenoid and plastoquinone biosynthetic pathways (Pallett et al., 1998).

metal ion at the catalytic center of the active enzyme is Fe^{2+}. In the enzyme-inhibitor complex, the three amino acids coordinating to the metal ion remain the same but two coordinating water molecules have been displaced by the 1,3-diketone moiety of the inhibitor DAS869. In addition to metal coordination, the inhibitor binding site involves the side chains of several residues, most notably the phenyl groups of Phe-360 and Phe-403, which form a π-stacking interaction with the benzoyl moiety of DAS869. The N1-*tert*-butyl

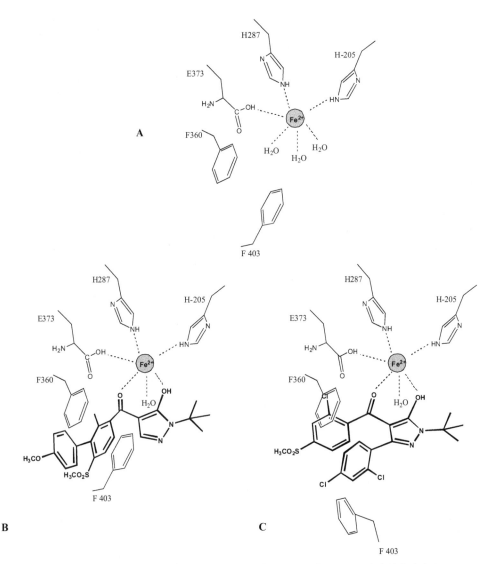

Fig. 7. Schematic representation of the active site of *At*HPPD (Yang et. al, 2004). (A) Active site of the enzyme without herbicidal substrate. (B) *At*HPPD-DAS869 complex. (C) *At*HPPD-DAS645 complex.

group on the ligand pyrazole has a tight fit against Pro-259 and causes a shift of ~0.5 Å compared to its position in uncomplexed *At*HPPD. No hydrogen bonding interactions with the inhibitor were detected. The structure of DAS645 a plant selective inhibitor in complex with *At*HPPD showed similar binding pattern as it was with DAS869 but with few notable differences. Because of the steric presence of the 3-(2,4-dichlorophenyl) substitution on the pyrazole, Phe-403 has rotated away from the inhibitor.

The interaction between the β-triketones and the catalytic site of *At*HPPD was modeled by docking of inhibitors into the active site plant HPPD (Dayan and Duke, 2007). The 1,3-diketone moiety of all the docked inhibitors coordinated Fe^{2+} ion still formed an octahedral complex with three strictly conserved active site residues (Glu-373, His-287 and His-205) and a critical binding site H_2O molecule, providing a strong ligand orientation and binding force. The observed interactions were consistent with those established with several classes of potent 1,3-diketone-type HPPD inhibitors. The β-triketone-rich essential oil of manuka (*Leptospermum scoparium*) and its components leptospermone, isoleptospermone, and grandiflorone were inhibitory to HPPD. Structure-activity relationhips indicated that the size and the lipophilicity of their side-chains affected the potency of the compounds. Both the the exceedingly tight association of HPPD inhibitorsand the relatively slow onset of inhibition are consistent with such inhibitors acting as transition state analogs (Kavana et al., 2003).

Identification of catalytic residues in active site of the Carrot HPPD protein has also been disclosed (Raspail et al., 2011). The results highlights a) the central role of Gln-272, Gln-286, and Gln-358 in HPP binding and the first nucleophilic attack, b) the important movement of the aromatic ring during the reaction, and c) the key role of Asn-261 and Ser-246 in C1 hydroxylation and the final ortho rearrangement steps (numbered according to *At*HPPD crystal structure).

2.3 Interaction of acetyl-CoA carboxylase (ACC) inhibitors with the target site enzyme

Acetyl-CoA carboxylases (ACCs) are crucial for the biosynthesis of fatty acids. They catalyze the production of malonyl-CoA from acetyl-CoA and CO_2, a reaction that also requires the hydrolysis of ATP (Shaner 2003) (Figure 8). Cyclohexanediones such as sethoxydim and the

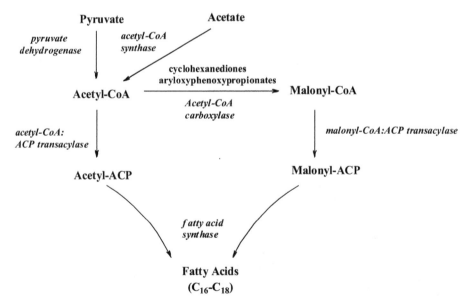

Fig. 8. Schematic representation of fatty acid biosynthesis.

aryloxyphenoxypropionates such as haloxyfop and diclofop, two different classes of widely used commercial herbicides are known inhibitors of ACCs (Burton, 1997). In grasses, such as wheat and maize, ACC is a high molecular weight, multi-domain enzyme, whereas in broadleaf species ACC exists as a multi-subunit enzyme. The cytosolic form of ACCs is a multi-subunit enzyme. The herbicidal ACC inhibitors specifically inhibit the multi-domain enzyme that is in the Gramineae and therefore they can be selectively used to control grasses in broadleaf crops. The molecular mechanism for the inhibition of the carboxyltransferase (CT) domain of ACC by haloxifop and diclofop herbicides was established by analyses of crystal structure of a complex of the yeast enzyme with the herbicides (Zhang et al., 2004). Haloxyfop is bound in the active site region, at the interface between the N domain of one monomer and the C domain of the other monomer of the dimer (Figure 9). The pyridyl ring of the inhibitor is sandwiched between the side-chains of

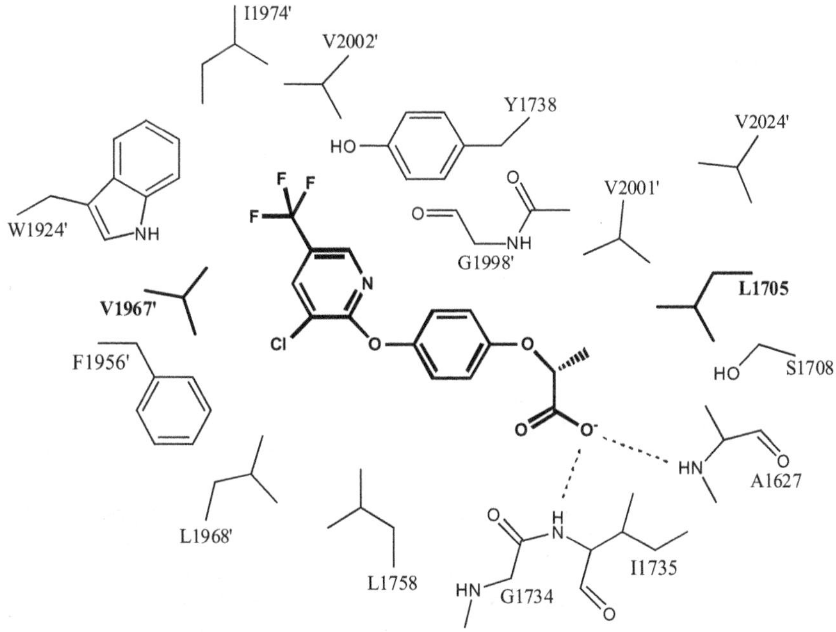

Fig. 9. Schematic representation of the interaction between haloxyfop and the CT domain (Zhang et al., 2004).

Tyr-1738 and Phe-1956′ (primed residue numbers indicate the C domain of the other monomer), showing π-π interaction. The trifluoro-methyl group is positioned over the plane of the Trp-1924′ side chain, as well as near the side-chains of Val-1967″, Ile-1974′, and Val-2002′. The phenyl ring in the center of the inhibitor is situated between the Gly-1734–Ile-1735 and Gly-1997′–Gly-1998′ amide bonds. One of the carboxylate oxygen atoms of the inhibitor is hydrogen-bonded to the main-chain amides of Ala-1627 and Ile-1735 whereas the other is exposed to the solvent. The methyl group of haloxyfop has van der Waals interactions with the side chains of Ala-1627 and Leu-1705. In contrast, this methyl group in the S stereoisomer of haloxyfop will clash with one of the carboxylate oxygens of the inhibitor explaining the selectivity for the R stereoisomer of this class of compounds. There are substantial conformational changes in the active site of the enzyme upon herbicide binding. Most importantly, the side chains of Tyr-1738 and Phe-1956' assume new positions in the inhibitor complex to become π-stacked with the pyridyl ring of haloxyfop. A similar binding pattern was shown for diclofop. Most of the residues that interact with the herbicides are either strictly or highly conserved. Only two residues show appreciable variation among the different CT domains, Leu-1705 and Val-1967'. Variation/mutation of these residues can confer resistance to the herbicides in plants (Zagnitko et al., 2001). The residue that is equivalent to Leu-1705 in the CT domains of wheat and other sensitive ACCs is Ile and the Ile→Leu mutation renders the enzyme resistant to both haloxyfop and sethoxydim. The residue that is equivalent to Val-1967'in sensitive plants is Ile and the Ile→Asn mutation makes the plants resistant to FOPs but not to the DIMs. The Ile→Val mutation can confer resistance to haloxifop and does not affect the sensitivity to clodinafop (Delye et al., 2003).

2.4 Interaction of auxin herbicides with auxin receptors

A plant hormon, naturally occurring auxin (indole-3-acetic acid, IAA) regulates plant growth by modulating gene expression leading to changes in cell division, elongation and differentiation (Woodward and Bartel, 2005). IAA coordinates many plant growth processes by modulating gene expression which leads to changes in cell division, elongation and differentiation. The perception of auxin signal by cells has been a topic of research. A receptor for auxin was identified as the F-box protein TIR1 (transport inhibitor response 1) is a component of cellular protein complex (SCF[TIR1]) (Tan et al., 2007). TIR1 was reported to recognize synthetic auxin analogues such as 1-naphthalene acetic acid (1-NAA) and 2,4-dichlorophenoxiacetic acid (2,4-D). Similarly to IAA, both compounds are able to promote the binding of Aux/IAA proteins to the TIR1 F-box protein. Auxin-induced genes are regulated by two classes of gene-transcription factors, auxin/response factors (ARFs) and the Aux/IAA repressors. The auxin signalling pathway includes ARFs binding to auxin-response promoter elements in auxin-response genes. When auxin concentrations are low, Aux/IAA repressors associate with ARF activators and repress the expression of the genes. When auxin concentrations increase, auxin binds to TIR1 receptor in the SCF[TIR1] complex, leading to recruitment of the Aux/IAA repressors to TIR1. Once recruited to SCF[TIR1] complex, the repressors enter a pathway that leads to their destruction and the subsequent activation of the auxin/response genes. A recently published crystalline structure of TIR1 complex showing how auxins fit into the surface pocket of TIR1 and enhance the binding Aux/IAA repressors to TIR1. In contrast to an allosteric mechanism, auxin binds to the same TIR1 pocket that docks the Aux/IAA substrate. Without inducing significant conformational changes in its receptor, auxin increases the affinity of two proteins by simultaneously interacting with both in the cavity at protein interface functioning as a 'molecular glue'. The synthetic auxins bind to TIR1 in a similar manner, but with affinities determined by how effectively their ring structures fit into and interact with the promiscuous cavity of the receptor.

2.5 Research for finding new target sites

Demand for new herbicides having with novel mode of action is a continual challenge stimulated by several reasons such as weed resistance evolved to several classes of herbicides as well as strict environmental and safety requirements.

Adenylosuccinate synthase (AdSS), an essential enzyme for *de novo* purine synthesis was found as a promising herbicide target site for hydantocidin (Siehl et al., 1996), a naturally occurring spironucleoside isolated from *Streptomyces hygroscopicus* (Haruyama et al., 1991). Hydantocidin was shown to be a proherbicide that, after phosphorylation at the 5' position, inhibits adenylosuccinate synthase (Fonné-Pfister et al., 1996). The mode of binding of hydantocidin 5'-monophosphate (HMP) to the target enzyme from *E. coli* was analyzed by determining the crystal structure of the enzyme inhibitor complex. The binding site was found at one end of a crevice across the middle of the enzyme. It was shown that AdSS binds the phosphorylated hydantocidin at the same site as it does adenosine 5'-monophosphate, the natural feedback regulator of this enzyme. The phosphate group is very important for binding to the enzyme and involves most of the direct contacts that the inhibitors have with the protein, including hydrogen bonds to Arg-143 from the other monomer in the dimer. The 2' and 3' hydroxyl groups of the ribose moiety have hydrogen bonds with Arg-303 and the main-chain carbonyls of Gly-127 and Val-273. The sugar groups have slightly different positions in the binding sites, which may be favorable in the case of HMP due to the internal hydrogen bond between the hydantoin and the phosphate groups. In the region where the structures of the two ligands differ, most of the contacts with the protein are made in both cases via water molecules. The hydantoin moiety is coplanar with the adenosine group and many of the water molecules lying in the same plane.

Embryonic factor 1 (*FAC1*) is one of the earliest expressed plant genes and encodes an AMP deaminase (AMPD), which was identified as a herbicide target (Dancer et al., 1997). Coformycin (Isaac et al., 1991) produced by various microbes and carbocyclic coformycin (Bush et al., 1993) isolated from a fermentation of *Saccharothrix* spp. have previously been reported to have herbicidal activity. AMPD catalyzes deamination of AMP to inosine-5'-monophosphate and together with the adenylosuccinate synthase (AdSS) and adenylosuccinate lyase it forms the purine nucleotide cycle. The N-terminal transmembrane domain in *Arabidopsis FAC1* was indentified using a recombinant enzyme (Han et al., 2006). The active site of FAC1 with bound coformycin 5'-phosphate, a herbicidally active form of coformycin, is positioned on the C-terminal side of the imperfect $(\alpha/\beta)_8$-barrel, surrounded by multiple helices and loops. The catalytic zinc ion is coordinated to the coformycin 5'-phosphate, an aspartic acid (Asp-736) and three histidine (His-391, 393, and 659) residues. The phosphate group of the inhibitor is located in a polar environment. The transmembrane domain and disordered linker domain of both subunits tether the dimeric globular catalytic domain to the lipid bilayer. The basic residue-rich surface spanning the dimer interface can become quite flat in the region of positive charge and facilitate interaction with negative patches on the surface of the membrane. However, the mechanistic bases for lethality associated with dramatic reductions in plant AMPD activity remain to be elucidated.

Inhibition of the enzymes of amino acid biosynthesis is an important target for several classes of herbicides as detailed earlier.

A new target, tryptophan synthase (TRPS) catalyzes the final two steps in the biosynthesis of tryptophan (Metha and Christen, 2000). It is typically found as an $\alpha2\beta2$ tetramer (Raboni et al., 2009). The α subunits catalyze the reversible formation of indole and glyceraldehyde-3-phosphate (G3P) from indole-3-glycerol phosphate (IGP). The β subunits catalyze the irreversible condensation of indole and serine to form tryptophan in a pyridoxal phosphate (PLP) dependent reaction. Each α active site is connected to a β active site by a 25 Å long hydrophobic channel contained within the enzyme. This facilitates the diffusion of indole formed at α active sites directly to β active sites in a process known as substrate channeling. A rational design of TRPS inhibitors as herbicides based on the structure of the inhibitory indole-3-phosphate (Rhee et al., 1998) in complex with the enzyme has been described (Sachpatzidis et al., 1999). Series of 4-aryl-thiobutylphosphonates were designed in which sulfur mimics the sp3-hybridized intermediate of the natural reaction intermediate and opening the heteroaromatic ring of the indole resulted in an increased rotational freedom. Amino group place in *ortho-postion* provided a potent inhibitory compound (IC_{50}= 178 nM) which was shown to bind to the α-site of the enzyme. Later it was shown that the indole-3-acetyl amino acids such as indole-3-acetylglycine and indole-3-acetyl-l-aspartic acid are both α-subunit inhibitors and β-subunit allosteric effectors, whereas indole-3-acetyl-l-valine is only an α-subunit inhibitor (Marabotti et. Al, 2000). The crystal structures of tryptophan synthase complexed with indole-3-acetylglycine and indole-3-acetyl-l-aspartic acid revealed that both ligands bind to the active site such that the carboxylate moiety is positioned similarly as the phosphate group of the natural substrates (Weyand et al., 2002).

Since biotin a cofactor for enzymes involved in carboxylation, trans-carboxylation and decarboxylation reactions dethiobiotin synthase (DTBS) can also be a promising target to develop new herbicides. DTBS is a penultimate enzyme in the biotin biosynthesis catalyzing the formation of a cyclic urea precursor of biotin from diaminopelargonic acid, CO_2 and ATP (Marquet et. al, 2001). Bacterial enzyme was used as a model for the dsign and synthesis of DTBS inhibitors as herbicides. (Rendina et al., 1999). In order to mimic the carbamate intermediate of the DTBS various carboxylates and phosphonates were prepared but poor level of inhibition thus weeak *in vivo* activities were detected. Co-crystallisation of a diphosphonate derivative with the enzyme revealed a relatively close binding to surface of the enzyme in a solvent exposed region which may explain the weak levels of inhibition. However no other reports were found on the synthesis of more potent inhibitors.

Pyruvate phosphate dikinase (PPDK) is an enzyme that catalyzes the inter-conversion of adenosine triphosphate (ATP), phosphate (Pi), and pyruvate with adenine monophosphate (AMP), pyrophosphate (PPi), and phosphoenolpyruvate (PEP) in the presence of magnesium and potassium/sodium ions ($Mg2+$ and $K+/Na+$) (Evans and Wood, 1968). The three-step reversible reaction proceeds via phosphoenzyme and pyrophosphoenzyme intermediates with a histidine residue serving as the phosphocarrier. The enzyme has been found in bacteria, in C_4 and Crassulacean acid metabolism plants, and in parasites, but not in higher animal forms. Inhibition of PPDK significantly hinders C_4 plant growth (Maroco et al., 1998). A total of 2,245 extracts from 449 marine fungi were screened against C_4 plant PPDKs as potential herbicide target (Motti et al., 2007). Extracts from several fungal isolates selectively inhibited PPDK. Bioassay-guided fractionation of one isolate led to the isolation of the known compound unguinol, which inhibited PPDK with a 50% inhibitory concentration of 42.3 μM. Unguinol had deleterious effects on a model C_4 plant but no effect on a model C_3 plant. The results indicated that unguinol inhibits PPDK via a novel mechanism of action which also translates to a herbicidal effect on whole plants.

Classical photosystem-II (PSII) inhibitors, such as urea, triazine and triazinone herbicides, bind to the D1 protein in stoichiometric fashion. Due to herbicide binding, the electron flow from PSII is disrupted and carbon dioxide fixation ceases. Since the electron acceptor is not able to accept electrons from photo-excited chlorophyll, free radicals are generated and chlorosis develops (Ahrens and Krieger-Liszkay, 2001). Stoichiometric inhibition of D1 as an herbicide mode of action has a major disadvantage since high application rates of the herbicide required for the activity. Much lower rates of herbicide would be necessary if the inhibition of the biosynthesis of mature D1 protein were targeted. A carboxyterminal processing protease (CptA) was chosen as a target and tested in a high throughput screen for CptA inhibitors (Duff et al., 2007). CptA, a low abundance enzyme located in the thylakoids catalyzes the conversion of the nascent pre-D1 protein into the active form of D1 by cleaving the 9 C-terminal residues. Under light conditions D1 protein is continously being damaged by light and is turned over. Thus the active CtpA is constantly required under light conditions to maintain fuctional PSII complexes. The herbicidal effects by the inhibition of CtpA protease were reported using novel CptA protease inhibitors. Altough K_i values of CtpA inhibitors were in micromolar range the *in planta* results suggested that good CptA inhibitors exhibited effective herbicidal activity while compounds with poor inhibitory activity possessed with only observable phytotoxicity.

3. Conclusions

Increasing problems associated with herbicide resistance as well as growing demand for herbicides in the developing world will expectedly spur the research and development for new herbicides. In the last decade, strategies for discovery of new herbicides have shifted from the testing of molecules in whole plant studies towards the use of *in-vitro* assays against molecular targets (Lein et al., 2004). 3D structures produced by X-ray crystallography have become an integral part of the current agrochemical and pharmaceutical discovery process. As genomic and proteomic data becomes increasingly available, a large numbers of validated targets will provide a basis for the structure-based inhibitor design (Walter, 2002) as a routine approach to obtain lead compounds.

4. References

Ahrens W.H. & Krieger-Liszkay A. (2001). Herbicide-induced oxidative stress in photosystem II. *Trends Biochem Sci.*, 26, 648-653.

Amrhein N, Deus B, Gehrke P, & Steinrucken HC (1980). The site of the inhibition of the shikimate pathway by glyphosate. II. Interference of glyphosate with chorismate formation *in vivo* and *in vitro*. *Plant Physiol.* 66, 830-834.

Ashigh J. & Tardif F. J. (2007). An Ala205Val substitution in acetohydroxyacid synthase of eastern black nightshade (*Solanum Ptychanthum*) reduces sensitivity to herbicides and feedback inhibition. *Weed Sci.*, 55, 558-565.

Barry G. F. & Kishore G. A. (1995). Glyphosate tolerant plants. US Patent 5,463,175.

Bernasconi P., Woodworth A. R., Rosen B. A., Subramanian M. V., & Siehl D. L. (1995). A naturally occurringpoint mutation confers broad range tolerance to herbicides that target acetolactate synthase. *J. Biol. Chem.*, 270, 17381-17385.

Brownlee J., Johnson-Winters K., Harrison D. H. T., & Moran G. R. (2004). Structure of the ferrous form of (4-hydroxyphenyl)pyruvate dioxygenase from Streptomyces avermitilis in complex with the therapeutic herbicide, NTBC. *Biochemistry*, 43, 6370-6377.

Burton J. D. (1997). Acetyl-coenzyme A carboxylase inhibitors. In: *Herbicide activity: Toxicology, biochemistry and molecular biology*, Roe R. M., Burton J. D., Kuhr R. J., pp 187–205, IOS Press, Washington, DC, USA.

Bush B. D., Fitchett G. V., Gates D. A., & Langley D. (1993). Carbocyclic nucleosides from a species of *Saccharothrix*. *Phytochemistry*, 32, 737-739.

Castle L. A., Siehl D. L., Gorton R., Patten P. A., Chen Y. H., Bertain S., Cho H. J., Duck N., Wong J., Liu D., & Lassner M. W. (2004). Discovery and directed evolution of a glyphosate tolerance gene. *Science*, 304, 1151-1154.

Dancer J. E., Hughes R. G., & Lindell S. D. (1997). Adenosine-5′-phosphate deaminase. A novel herbicide target. *Plant. Physiol.*, 114, 119-129.

Delye C., Zhang X.-Q., Chalopin C., Michel S., & Powles S. (2003). An isoleucine residue within the carboxyl-transferase domain of multidomain acetyl-coenzyme A carboxylase is a major determinant of sensitivity to aryloxyphenoxypropionate but not to cyclohexanedione inhibitors *Plant Physiol.*, 132, 1716-1723.

Dayan F. E, Duke S. O., Sauldubois A., Singh N., McCurdy C., Cantrell C. (2007). p-Hydroxyphenylpyruvate dioxygenase is a herbicidal target site for β-triketones from Leptospermum scoparium. *Phytochemistry*, 68, 2004-2014.

Duff S. M. G., Chen Y-C. S., Fabbri B. J., Yalamanchili G., Hamper B. C., Walker D. M., Brookfield F. A., Boyd E. A., Ashton M. R., Yarnold C. J., & CaJacob C. A. (2007). The carboxyterminal processing protease of D1 protein: Herbicidal activity of novel inhibitors of the recombinant and native spinach enzymes. *Pest. Biochem. Physiol.*, 88, 1-13.

Duggleby R. G. & Pang S. S. (2000). Acetohydroxyacid synthase. *J. Biochem. Mol. Biol.*, 33, 1-36.

Evans H. J. & Wood H. G. (1968. The mechanism of the pyruvate, phosphate dikinase reaction. *Proc. Natl. Acad. Sci. USA*. 61, 1448-1453.

Fonné-Pfister R., Chemla P., Ward E., Girardet M., Kreuz K. E., Honzatko R. B., Fromm H. J., Schär H-P., Grütter M. G., & Cowan-Jacob S. W. (1996). The mode of action and the structure of a herbicide in complex with its target: Binding of activated hydantocidin to the feedback regulation site of adenylosuccinate synthetase. *Proc. Natl. Acad. Sci. USA*, 93, 9431-9436.

Franz J. E., Mao M. K., & Sikorski, J. A. (1997). *Glyphosate: A Unique Global Herbicide*, ACS Monograph Series, No. 189, ACS, Washington, DC.

Fritze I. M., Linden L., Freigang J., Auerbach G., Huber R., & Steinbacher S. (2004). The crystal structures of Zea mays and Arabidopsis 4-Hydroxyphenylpyruvate Dioxygenase. *Plant Physiol.*, 134, 1388-1400.

Funke T., Han H., Healy-Fried M. L., Fischer M., & Schönbrunn E. (2006). Molecular basis for the resistance of Roundup Ready crops. *Proc. Natl. Acad. Sci. USA*, 103, 13010-13015.

Funke T., Yang Y., Han H., Healy-Fried M., Olensen S., Becker A., & Schönbrunn E. (2009). Structural basis of glyphosate resistance resulting from the double mutation Thr[97]→Ile and Pro[101]→Ser in 5-enolpyruvylshikimate-3-phosphate synthase from *Escherichia coli*. *J. Biol. Chem.*, 284, 9854-9860.

Gerwick B. C., Subramanian M. V., Loney-Gallant V., & Chandler D. P. (1990). Mechanism of action of the 1,2,4-triazolo[1,5-α]pyrimidine. *Pestic. Sci.*, 29, 357–364.

Han B. W., Bingman C. A., Mahnke D. K., Bannen R. M., Bednarek S. Y., Sabina R. L., Phillips, G. N. Jr. (2006). Membrane association, mechanism of action, and structure of *Arabidopsis* embryonic factor 1 (*FAC1*). *J. Biol. Chem.*, 281, 14939-14947.

Haruyama H., Takayama T., Kinoshita T., Kondo M., Nakajima M., & Haneishi T. (1991). Structural elucidation and solution conformation of the novel herbicide hydantocidin. *J. Chem. Soc. Perkin Trans.* 1, 1637-1640.

Hattory J., Rutledge R., Labbe H., Brown D., Sunohara G., & Miki B. (1992). Multiple resistance to sulfonylureas and imidazolinones conferred by an acetohydroxyacid synthase gene with separate mutations for selective resistance. *Mol. Gen. Genet.*, 232, 167-173.

Isaac B. G., Ayer S. W., Letendre L. J., & Stonard R. J. (1991). Herbicidal nucleosides from microbial sources. *J. Antibiotics*, 44, 729-732.

Job V., Molla G., Pilone M. S., & Pollegioni L. (2002). Overexpression of a recombinant wild-type and His-tagged *Bacillus subtilis* glycine oxidase in *Escherichia coli. Eur J. Biochem.*, 269, 1456-1463.

Kavana M. & Moran G. R. (2003). Interaction of (4-hydroxyphenyl)pyruvate dioxygenase with specific inhibitor 2-[2-nitro-4-(trifluoromethyl)benzoyl]-1,3-cyclohexadione. *Biochemistry*, 42, 10238-10245.

Klausener A., Raming K., & Stenzel K. (2007) Modern tools for drug discovery in agricultural research, In: *Pesticide Chemistry. Crop Protection Public Health, Environmental safety.* Ohkawa H., Miyagawa H., Lee P. W., pp 55-63, Wiley-VCH, Weinheim.

LaRossa R. A. & Schloss J. V. (1984). The sulfonylurea herbicide sulfometuron is an extremely potent and selective inhibitor of acetolactate synthase in Salmonella typhimurium. *J. Biol. Chem.*, 259, 8753-8757.

Le D. T., Yoon M-Y., Kim Y. T., Choi J.-D. (2005). Roles of three well-conserved argine residues in mediating the catalytic activity of tobacco acetohydroxy acid synthase. *J. Biochem.*, 138, 35-40.

Lee H., Rustgi S., Kumar N., Burke I., Yenish J. P., Kulvinder S. G., von Wettstein D., & Ullrich S. E. (2011). Single nucleotide mutation in the barley acetohydroxyacid synthase (AHAS) confers resistance to imidazolinone herbicides. *Proc. Natl. Acad. Sci. USA*, 108, 8909-8913.

Lein W. F., Börnke F., Reindl A., Erhardt T., Stitt M., & Sonnewald U. (2004). Target-based discovery of novel herbicides. *Curr. Opin. Plant Biol.*, 7, 219-225.

Lydon J. & Duke S. O. (1999). Inhibitors of glutamine biosynthesis, In *Plant amino acids: biochemistry and biotechnology*, Singh B. K., pp 445–464, Marcel Dekker, New York, USA.

Marabotti A., Cozzini P., & Mozzarelli A. (2000). Novel allosteric ejectors of the tryptophan synthase K2L2 complex identified by computer-assisted molecular modeling. *Biochim. Biophys. Acta*, 1476, 287–299.

Maroco J. P., Ku M. S. B., Lea P. J., Dever L. V., Leegood R. C., Furbank R. T., & Edwards G. E.. (1998). Oxygen requirement and inhibition of C4 photosynthesis—an analysis of C4 plants deficient in the C3 and C4 cycles. *Plant Physiol.* 116, 823–832.

Marquet A., Bui B. T., & Florentin D. (2001). Biosynthesis of biotin and lipoic acid.*Vitam. Horm.*, 61, 51-101.

McCourt J. A., Pang S. S., King-Scott J., Guddat L.W., & Duggleby R. G. (2006). Herbicide-binding sites revealed in the structure of plant acetohydroxyacid synthase. *Proc. Natl. Acad. Sci. USA*, 103, 569-573.

Metha P. K. & Christen P. (2000). The molecular evolution of pyridoxal-5'-phosphate-dependent enzymes. *Adv. Enzymol. Relat. Areas Mol. Biol.*, 74, 129-184.

Mohamed Naseer Ali M., Kaliannan P., & Venuvanalingam P. (2005). *Ab initio* computational modeling of glyphosate analogs: Conformational perspective. *Struct. Chem.*, 16, 491-506.

Moran G. R. (2005). 4-Hydroxyphenylpyruvate dioxygenase. *Arch. Biochem. Biophys.*, 433, 117-128.

Motti C. A., Bourne D. G., Burnell J. N., Doyle J. R., Haines D. S., Liptrot C. H., Llewellyn L. E., Ludke S., Muirhead A., & Tapiolas D. M. (2007). *Appl. Environ. Microbiol.*, 1921-1927.

Padgette S. R., Kolacz K. H., Delannay X., Re D. B., LaVallee B. J., Tinius C. N., Rhodes W. K., Otero Y. I., Barry G. F., Eichholz D. A., Peschke V. M., Nida D. L., Taylor N. B., & Kishore G. M. (1995). Development, identification, and characterization of a glyphosate-tolerant soybean line. *Crop Sci.*, 35, 1451-1461.

Pallett K. E., Little J. P., Sheekey M., Veerasekaran P. (1998). The mode of action of isoxaflutol. I. Physiological effects, metabolism, and selectivity. *Pestic. Biochem. Physiol.*, 62, 113-124

Pang S.S., Guddat L. W., & Duggleby R. G. (2004). Crystallization of Arabidopsis thaliana acetohydroxyacid synthase in complex with the sulfonylurea herbicide chlorimuron ethyl. *Acta Cryst. D*, 60, 153-155.

Pascal R. A., Oliver M. A., & Chen, Y.-C. J. (1985). Alternate substrates and inhibitors of bacterial 4-hydroxyphenylpyruvate dioxygenase. *Biochemistry*, 24, 3158-3165.

Perez-Jones A., Park K.-W., Polge N., Colquhoun J., Mallory-Smith C. A. (2007). Investigating the mechanisms of glyphosate resistance in *Lolium multiflorum. Planta*, 226, 395-404.

PedottiM., Rosini E., Molla G., Moschetti T., Savino C., Vallone B., & Pollegioni L. (2009). Glyphosate resistance by engineering the flavoenzyme glycine oxidase. *J. Biol. Chem.*, 284, 36415-36423.

Pollegioni L., Schönbrunn E., & Siehl D., (2011). Molecular basis of glyphosate resistance – different approaches through protein engineering. *FEBS J.*, 278, 2753-2766.

Prisbylla M. P., Onisko B. C., Shribbs J. M., Adams D. O., Liu Y., Ellis M. K., Hawkes T. R., & Mutter L. C. (1993). The novel mechanism of action of herbicidal triketones, *Proc. British Crop Prot. Conf. – Weeds*, vol. 2, 731-738.

Raboni S., Bettati S., & Mozzarelli A. (2009). Tryptophan synthase: a mine for enzymologists. *Cell. Mol. Life Sci.*, 66, 2391-2403.

Raspail C., Graindorge M., Moreau Y., Crouzy S., Lefevre B., Robin A. Y., Dumas R., & Matringe M. (2011). 4-Hydroxyphenylpyruvate dioxigenase catalysis: Identification of catalytic residues and production of a hydroxylated intermediate shared with a structurally unrelated enzyme. *J. Biol. Chem.*, http://www.jbc.org/cgi/doi/10.1074/jbc.M111.227595.

Rendina A. R., Taylor W. S., Gibson K., Lorimer G., Rayner D., Lockett B., Kranis K., Wexler B., Marcovici-Mizrahi D., Nudelman A., Nudelman A., Marsilii E., Chi H., Wawrzak Z., Calabrese J., Huang W., Jia J., Schneider G., Lindqvist Y., & Yang G. (1999). The design and synthesis of inhibitors of dethiobiotin synthetase as potential herbicid. *Pestic. Sci.*, 55, 236-247.

Rhee S., Miles E. W., & Davies D. R. (1998). Cryo-crystallography of a true substrate, indole-3-glycerol phosphate, bound to a mutant (d60n) tryptophan synthase 22 complex reveals the correct orientation of active site Glu49. *J. Biol. Chem.*, 263, 8611-8614.

Rost T. L. (1984). The comparative cell cycle and metabolic effects of chemical treatments on root tip meristems. *J. Plant Growth Regul.*, 3, 51-63.

Sachpatzidis A., Dealwis C., Lubetsky J. B., Liang P.-H., Anderson K. S., & Lolis E. (1999). *Biochemistry*, 38, 12665-12674. Crystallographic studies of phosphonate-based alpha-reaction transition-state analogues complexed to tryptophan synthase.

Schönbrunn E., Eschenburg S., Shuttleworth W. A., Schloss J. V., Amrhein N., Evans J. N. S., & Kabsch W. (2001) Interaction of the herbicide glyphosate with its target enzyme 5-enolpyruvylshikimate 3-phosphate synthase in atomic detail. *Proc. Natl. Acad. Sci. USA*, 98, 1376-1380.

Serre L., Sailland D. S., Boudec P., Rolland A., Pebay-Peyroula E., & Cohen-Addad C. (1999). Crystal structure of Pseudomonas fluorescens 4-hydroxyphenylpyruvate dioxygenase: an enzyme involved in the tyrosine degradation pathway. *Structure*, 7, 977-988.

Shah D. D., Conrad J. A., Heinz B., Brownlee J. M., & Moran G. R. (2011). Evidence for the mechanism of hydroxylation by 4-hydroxyphenylpyruvate dioxygenase and hydroxymandelate synthase from intermediate partitionin in active site variants. *Biochemistry*, dx.doi.org/10.1021/bi2009344.

Shaner D. L., Anderson P. C., Stidham M. A. (1984). Imidazolinones: potent inhibitors of acetohydroxy acid synthase. *Plant Physiol.*, 76, 545-546.

Shaner D. L. (2003). Herbicide safety relative to common targets in plants and mammals. *Pest. Manag. Sci.* 60, 17-24.

Shaner D. L. & Singh B. J. (1993). Phytotoxicity of acetohydroxyacid synthase inhibitors is not due to accumulation of 2-ketobutyrate and/or 2-aminobutyrate. *Plant Physiol.*, 103, 1221-1226.

Shuttleworth W. A., Pohl M. E., Helms G. L., Jakeman D. L., & Evans J. N. S. (1999). Site-directed mutagenesis of putative active site residues of 5-enolpyruvylshikimate-3-phosphate synthase. *Biochemistry*, 38, 296-302.

Sibony M., Michael A., Haas H. U., Rubin B., & Hurle K. (2001). Sulfometuron-resistant Amaranthus retroflexus: Cross-resistance and molecular basis for resistance to acetolactate synthase (ALS) inhibiting herbicides. *Weed Res.*, 41, 509-522.

Siehl D. L., Subramanian M. V., Walters E. W., Lee S. F., Anderson R. J., & Toschi A. G. (1996). Adenylosuccinate synthetase: Site of action of hydantocidin, a microbial phytotoxin. *Plant Physiol.*, 110, 753-758.

Siehl D. L., Castle L. A., Gorton R., & Keenan R. J. (2007). The molecular basis of glyphosate resistance by an optimized microbial acetyltransferase. *J. Biol. Chem.*, 282, 11446-11455.

Sikorski J. A. & Gruys K. J. (1997). Understanding glyphosate's mode of action with EPSP synthase: Evidence favoring an allosteric inhibitor model. *Acc. Chem. Res.*, 30, 2-8.

Sost D. & Amrhein N. (1990). Substitution of Gly-96 to Ala in the 5-enolpyruvylshikimate-3-phosphate synthase of Klebsiella pneumoniae results in a greatly reduced affinity for the herbicide glyphosate. *Arch. Biochem. Biophys.*, 282, 433-436.

Stallings W. C., Abdel-Meguid S. S., Lim L. W., Shieh H. S., Dayringer H. E., Leimgruber N. K., Stegemann R. A., Anderson K. S., Sikorski J. A., Padgette S. R., & Kishore G. M.

(1991). Structure and topological symmetry of the glyphosate target 5-enolpyruvylshikimate-3-phosphate synthase: a distinctive protein fold. *Proc. Natl. Acad. Sci. USA*, 88, 5046-5050.

Tan X., Calderon-Villalobos L. I. A., Sharon M., Zheng C., Robinson C.V., Estelle M., & Zheng, N. (2007). Mechanism of auxin perception by TIR1 ubiqutin ligase. *Nature*, 446, 640-645.

Tranel P. J., Wright T. R. (2002). Resistance of weeds to ALS-inhibiting herbicides: What have we learned. *Weed Sci.*, 50, 700-712.

Unno H., Uchida T., Sugawara H., Kurisu G., Sugiyama T., Yamaya T., Sakakibara H., Hase T., & Kusunoki M. (2006). Atomic structure of plant glutamine synthetase. A key enzyme for plant productivity. *J. Biol. Chem.*, 281, 29287-29296.

Yang C., Pflugrath J. W., Camper D. L., Foster M. L., Pernich D. J., & Walsh T.A. (2004). Structural basis for herbicidal inhibitor selectivity revealed by comparison of crystal structures of plant and mammalian 4-hydroxyphenylpyruvate dioxygenases. *Biochemistry*, 43, 10414–10423.

Yu Q., Cairns A., & Powles S. (2007). Glyphosate, paraquat and ACCase multiple herbicide resistance evolved in a *Lolium rigidum* biotype. *Planta*, 225, 499-513.

Walter M. W. (2002). Structure-based design of agrochemicals. *Nat. Prod. Rep.*, 19, 278-291.

Waltz E. (2010). Glyphosate resistance threatens Roundup hegemony. *Nat. Biotechnol.*, 28, 537-538.

Weyand M., SCHlichting I., Marabotti A., & Mozzarelli A. (2002). Crystal structures of a new class of allosteric effectors complexed to tryptophan synthase. *J. Biol. Chem.*, 277, 10647-10652.

Woodward A. W. & Bartel B. (2005). Auxin: regulation, action, and interaction. *Ann. Bot. (Lond.)*. 95, 707-735.

Zagnitko O., Jelenska J., Tevzadze G., Haselkorn I., & Gornicki P. (2001). An isoleucine/leucine residue in the carboxyltransferase domain of acetyl-CoA carboxylase is critical for interaction with aryloxyphenoxypropionate and cyclohexanedione inhibitors. *Proc. Natl. Acad. Sci. USA*, 98, 6617-6622.

Zhang H., Tweel B., & Tong L. (2004) Molecular basis for the inhibition of the carboxyltransferase domain of acetyl-coenzyme-A carboxylase by haloxyfop and diclofop. *Proc. Natl. Acad. Sci. USA*, 101, 5910-5915.

Zhou Q., Liu W., Zhang Y., & Liu K. K. (2007). Action mechanisms of acetolactate synthase-inhibiting herbicides. *Pestic. Biochem. Physiol.*, 89, 89-96.

Zimdahl R. L. (2002). My view. *Weed Sci.*, 50, 687.

2

Laboratory Study to Investigate the Response of *Cucumis sativus* L. to Roundup and Basta Applied to the Rooting Medium

Elżbieta Sacała, Anna Demczuk and Edward Grzyś
Wrocław University of Environmental and Life Sciences
Poland

1. Introduction

Roundup and Basta are commonly used herbicides in both agricultural and non-agricultural systems. They are non-selective broad spectrum post-emergence herbicides, relatively non-toxic to environment. Their mode of action is connected with inhibition of amino acids biosynthesis (Cobb, 1992; Wakabayashi & Böger, 2004). Herbicide Basta is also know as glufosinate or phosphinothricin. Glufosinate is a structural analogue of glutamic acid and it irreversibly inhibits glutamine synthetase (GS) which synthesises glutamine from glutamate and ammonium. Consequently, ammonium concentration in plant tissues strongly increases and causes metabolic disruption and plants' death. Roundup's active ingredient is glyphosate [N-(phosphonomethyl)glycine] that inhibits 5-enolpyruvyl shikimate-3-phosphate synthase (EPSPS). As a consequence, the biosynthesis of aromatic amino acids is inhibited. Besides, glyphosate is known to have secondary mode of action connected with the inhibition of chlorophyll biosynthesis resulting from a glyphosate-mediated inhibition of δ-aminolevulinic acid synthesis (Pline et al., 1999). Both herbicides alter the cellular pools of amino acids that results in disruption of nitrogen metabolism. Indirectly they affect protein synthesis, nitrate assimilation and nitrogen fixation (Bellaloui et al., 2006; Sacała et al., 2008). Nitrogen is an essential element that strongly influence and regulate plants' growth. For most plants in moderate climate regions, the main source of nitrogen is nitrate obtained from the soil. Nitrate is utilized in a linear pathway that involves its uptake and transport within the plant and then its assimilation, ammonium assimilation and amino acid and protein biosynthesis (Sitt et al., 2002). The first step in nitrate assimilation pathway, the reduction of nitrate to nitrite, is catalyzed by nitrate reductase (NR, EC 1.6.6.1–2). This enzyme plays a central role in nitrogen assimilation and is considered as a key point of metabolic regulation, as well as a rate-limiting enzyme in this pathway. Nitrate reductase is known to modulate rapidly to environmental stress conditions and is responsive to metabolic and physiological status of plants (Kaiser & Huber, 1994). Hence, sometime it is used as a biomarker of plant stress. Nitrate reduction occurs in both shoots and roots of plants and the relative contribution of each is dependent on the particular species, plant age and growth conditions.

Basta and Roundup are taken up by plants after application on the leaves and then are transported throughout the whole plant. Phytotoxicity of both herbicides is certainly rapid and leaf chlorosis, desiccation and necrosis may be observed within few days after application (Cobb, 1992).

In general opinion, Basta and Roundup don't accumulate in the environment and rapidly break down. However, there are reports showing that the half-life values of glyphosate vary on a wide range from a few days to several months and even years (Vereecken , 2005). Herbicide contamination in the soil may originate from foliar washing off, undirected spray drift, exudation from roots, and death and decomposition of treated plant residues (Hanke et al., 2010; Tesfamariam et al., 2009). Glyphosate is often found in surface waters and rarer in groundwater. For this reason, there is a risk that contamination of herbicide in soil may have detrimental effect on non-target plants.

The purpose of this study was to evaluate the effect of relatively low, non-lethal concentration of glyphosate and glufosinate, applied directly to root zone, on young fast growing cucumber seedlings. The concentrations of herbicides chosen in the assessment were not the same because a negative impact of examined herbicides on the environment and a risk of pollution are substantially different. Roundup is one of the most widely used herbicide in the world and its application is increasing all the time. This increase is due to a widespread cultivation of glyphosate-resistant crops (e.g. cotton, canola, corn, alfalfa, sugar beet). Hence, an expected amount of Roundup residues in soil is higher than that of Basta (Service, 2007; Woodburn 2000).

2. Experimental procedures

2.1 Plant material and growth conditions

The experiments were conducted on cucumber seedlings (*Cucumis sativus* L. var. Władko F-1) grown in hydroponic cultures. The experimental design consisted of following treatments: (i) control – 0.33-strength Hoagland nutrient solution, (ii) glyphosate treatment – 0.33-strength Hoagland nutrient solution plus Roundup herbicide in formulation 360 g a.i. per litre (at the concentration of 22 µl ·litre^{-1} of nutrient solution that corresponds to 0.0467 mM glyphosate), (iii) glufosinate treatment – control plus Basta herbicide in formulation 150 g a.i. per litre (at the concentration of 0.0077 mM glufosinate). Culture conditions were as follows: 16 h photoperiod (220 µmol·m^{-2}·s^{-1}) at 26/20°C day/night, 65-70% relative humidity.

2.2 Plant analyses

After 7 days of cultivation there were investigated growth parameters (length, fresh and dry weight of roots and shoots), activity of nitrate reductase (NRA) and concentrations of photosynthetic pigments and protein (soluble, insoluble and total).

2.2.1 Plant growth analysis

After 7 days of cultivation, plants were harvested and separated into roots and shoots and their lengths were measured. After that, plant organs (separately roots, hypocotyls and cotyledons) were weighted and dried at 105°C for 1 h and subsequently at 75°C. Then the dry weight was determined.

2.2.2 Determination of water content

The water content (WC) in roots and shoots was calculated on a fresh weight (FW) basis. WC was computed as the difference between the fresh weight and the dry weight divided by the fresh weight and multiplied by 100%.

2.2.3 Biochemical analyses

2.2.3.1 Photosynthetic pigments

Photosynthetic pigments were extracted from cotyledons using 80% acetone. Absorbance of obtained extracts was recorded at 470, 647, 663 nm and concentration of chlorophyll a, chlorophyll b, total chlorophyll (chl a + chl b) and carotenoids were calculated using the equations of Lichtenthaler (1987).

2.2.3.2 Protein concentration

For protein determination, 0.5 g of roots or cotyledons was homogenised with 0.1 M phosphate buffer (0.1 M K_2HPO_4-KH_2PO_4, pH 7.5 contained 1 mM EDTA) and then homogenate was centrifuged at 4°C (12000×g, 10 min). After centrifugation both the supernatant and the precipitate were collected. Soluble protein was measured in the supernatant according to the Lowry method (Lowry et al., 1951). Insoluble protein in the precipitate was extracted with 0.1 M NaOH at 60°C for 40 min and then the sample was centrifuged (15000×g, 10 min) and in supernatant protein was measured by the Lowry method.

2.2.3.3 Nitrate reductase activity (NRA)

Nitrate reductase activity was determined *in vivo* method according to Jaworski (1971). The plant material (1 g) was cut into 3 mm segments and placed in test-vial containing 20 cm³ phosphate buffer (0.1 M K_2HPO_4-KH_2PO_4, pH 7.5 contained 100 mM KNO_3 and 0.5% isopropanol). The test-vials were vacuum infiltrated for 2 min and then incubated in the dark at 25°C for 1 h. The amount of nitrite formed during the reaction was measured spectrophotometrically at 540 nm after adding 1% sulfanilamide in 0.1 M HCl and 0.02% N-naphtyl-ethylenediamine.

2.3 Statistical analysis

The experiment was arranged in a randomized complete block design and was repeated six times. The data for all parameters were statistically analysed using the variance analysis and the differences among mean values were compared by the least significant difference test (LSD, P≤0.05).

3. Results

3.1 Growth parameters and water content

All plants survived under examined conditions, however both herbicides caused significant reduction in fresh and dry weights of the above- and underground parts of plants (Fig. 1 A, B, C).

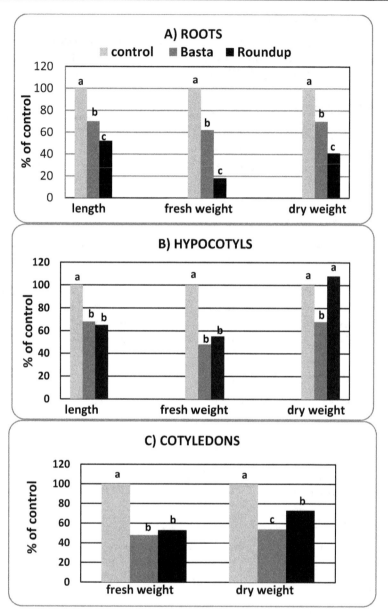

Fig. 1. The influence of Basta and Roundup on growth parameters of 7-day old cucumber seedlings. Values with the same letter above the bars do not differ significantly at P≤0.05.

The exception were hypocotyls of plants treated by Roundup. Their dry weight was similar to the control plants in contrast to fresh weight that was reduced by 45% comparing to non-treated plants. Roundup caused dramatic decrease in fresh weight of cucumber roots and this parameter lowered to 18% of the control value (Fig. 1A). Length of roots and their dry weight were also significantly reduced by Roundup treatment but this reduction did not exceed 60%. Basta also markedly inhibited growth of cucumber roots but this inhibition ranged from 30% for roots' length and dry weight to 38% for fresh weight. Basta and Roundup similarly affected height and fresh weight of hypocotyls whereas in dry matter accumulation there was difference (Fig. 1B). Basta caused 32% decrease in this parameter comparing to the control plants while Roundup did not act negatively. Both herbicides caused similar decrease in fresh weight of cotyledons (approximately 50% compared to the control plants) but Basta more reduced dry matter accumulation than Roundup (Fig. 1C).

Exposure of maize seedlings to herbicides had significant effect on water content in plants' tissues (Tab. 1).

	Water content (% of FW)		
	Control	Basta	Roundup
Roots	96.68±0.12 a	96.51±0.13 a	93.72±0.41 b
$LSD_{0.05}$	0.41		
Hypocotyls	96.46±0.26 a	95.49±0.38 b	92.08±0.20 c
$LSD_{0.05}$	0.46		
Cotyledons	92.81±0.17 a	92.07±0.27 b	89.21±0.24 c
$LSD_{0.05}$	0.37		

Table 1. The influence of Basta and Roundup on water content in different organs of 7-day old cucumber seedlings. Values in the same row followed by the same letter do not differ significantly at P≤0.05.

Roundup caused significant decrease in water content in all cucumber organs and the highest decline occurred in hypocotyls and it amounted to above 4% in comparison to the control plants. In cotyledons and roots, recorded decreases were respectively 3.60 and 2.96% compared to non-treated plants. Basta at examined concentration did not change significantly water content in cucumber roots but lowered this parameter significantly in hypocotyls and cotyledons (Tab. 1). However, observed changes in water content were considerably lower than under Roundup treatment (they did not exceed 1% comparing to the control plants). Hence, Roundup more effectively than Basta (in tested concentration) acted as a plant desiccant.

3.2 Photosynthetic pigments

Cotyledons of cucumber seedlings growing in nutrient medium contained herbicides were dark green. Chlorophyll concentration in these cotyledons was significantly higher than in the control plants (Tab. 2). Exposure of seedlings to Basta herbicide resulted in 8% increase in all forms of chlorophyll, while Roundup caused approximately 12% increase. Moreover, chlorophyll a to chlorophyll b ratio (Chl a/Chl b) was not affected and this value amounted about 3.0 under all treatments. Nevertheless, there were recorded differences in values of total chlorophylls to carotenoids ratio (Tab. 2). Under herbicides treatment this value showed tendency to lowering compared to non-treated plants. This change results from relatively higher increase in carotenoids contents comparing to increased level of chlorophylls. Basta and Roundup caused respectively 14 and 23% increase in carotenoids concentration comparing to the control plants.

	Control	Basta	Roundup
Chl a [$\mu g \cdot g^{-1}$ FW]	1134.2±37.5 b (100%)	1224.8±62.2 a (108%)	1267.8±8.2 a (112%)
LSD$_{0.05}$	63.3		
Chl b [$\mu g \cdot g^{-1}$ FW]	379.6±15.4 b (100%)	408.1±10.3 a (108%)	420.4±3.3 a (111%)
LSD$_{0.05}$	22.5		
Total Chl [$\mu g \cdot g^{-1}$ FW]	1513.8±51.5 b (100%)	1632.9±80.7a (108%)	1688.1±53.0 a (112%)
LSD$_{0.05}$	99.3		
Carotenoids [$\mu g \cdot g^{-1}$ FW]	239.1±2.7c (100%)	271.7±13.8 b (114%)	293.7±16.9 a (123%)
LSD$_{0.05}$	19.1		
Chl a/Chl b	2.99	3.00	3.02
Total Chl/Car	6.33	6.01	5.75

Table 2. Photosynthetic pigments concentration in cotyledons of 7-day old cucumber seedlings. Values in the same raw marked with the same letters don't differ significantly at P ≤ 0.05. Values in parentheses indicate the percent of control.

3.3 Protein concentration

Protein concentration in roots and cotyledons was affected differentially by examined herbicides (Tab. 3). Roundup caused significant increase in the protein concentration in both roots and cotyledons and in roots the total protein content increased by 42% in comparison to the control plants (Tab. 3). There was observed huge increase in soluble fraction of protein and it amounted to 166% of the value in the control plants. In cotyledons protein content was not such strongly affected as in roots, nevertheless insoluble fraction of protein was higher (15% increase) than in non-treated plants. Under Basta application there was observed a tendency to lowering protein concentration but the differences were not statistically significant.

3.4 Nitrate reductase activity (NRA)

Nitrate reductase activity was dramatically inhibited in cotyledons of cucumber growing in nutrient solution contained Roundup (Tab. 4). Enzyme activity lowered to 3% of the value in control plants. Basta application also caused decrease in enzyme activity (39% reduction) but it was relatively small compared to Roundup treatment. Both Basta and Roundup caused similar (approximately 2-fold) increase in nitrate reductase activity in cucumber roots.

	Protein concentration [mg·g^{-1} FW]		
	Control	Basta	Roundup
Roots			
Soluble protein	5.25±0.27 b (100%)	5.50±0.35 b (105%)	8.71±0.91 a (166%)
LSD$_{0.05}$	0.80		
Insoluble protein	4.76±0.46 b (100%)	4.20±0.53 b (88%)	5.51±0.48 a (116%)
LSD$_{0.05}$	0.70		
Total protein	10.01±0.55 b (100%)	9.70±0.58 b (97%)	14.22±0.90 a (142%)
LSD$_{0.05}$	1.06		
Cotyledons			
Soluble protein	19.60±1.19 a (100%)	19.64±1.29 a (100%)	21.70±2.07 a (111%)
LSD$_{0.05}$	2.70		
Insoluble protein	11.55±0.91 b (100%)	10.15±0.67 b (88%)	13.30±1.22 a (115%)
LSD$_{0.05}$	1.54		
Total protein	31.15±1.37 b (100%)	29.75±1.79 b (96%)	35.00±2.56 a (112%)
LSD$_{0.05}$	3.22		

Table 3. Protein concentration in roots and cotyledons of 7-day old cucumber seedlings. Values in the same raw marked with the same letters don't differ significantly at P ≤ 0.05. Values in parentheses indicate the percent of control.

	NRA [nmol(NO$_2^-$)·g^{-1} FW ·h^{-1}]		
	Control	Basta	Roundup
Roots	111.0±19.7 b (100%)	201.2±32.5 a (181%)	233.1±37.6 a (210%)
LSD$_{0.05}$	40.2		
Cotyledons	847.2±116.6 a (100%)	515.0±108.1 b (61%)	29.7±6.7 c (3%)
LSD$_{0.05}$	122.4		

Table 4. Nitrate reductase activity in roots and cotyledons of 7-day old cucumber seedlings. Values in the same raw marked with the same letters don't differ significantly at P ≤ 0.05. Values in parentheses indicate the percent of control.

4. Discussion

Herbicides are increasingly used in both agricultural and non-agricultural system as a quick, easy and inexpensive remedy for weeds control. There are many compounds registered as herbicides, which may be classified according to their mode of action. Very important and popular set of herbicides are that interfering with amino acid biosynthesis. They inhibit particular enzymes in plants and consequently block amino acid synthesis (Cobb, 1992; Tan et al., 2006). There are three major enzymes that can be inhibited: (1) acetolactate synthase (AHAS, EC 4.1.3.18) involved in the branched-chain amino acid biosynthesis pathway; (2) synthase of 5-enolpyruvyl shikimate-3-phosphate (EPSPS, EC 2.5.1.19) operating in the shikimate pathway and biosynthesis of aromatic amino acids; (3) glutamine synthetase (GS, EC 6.3.1.2) the key enzyme in ammonium assimilation. In our study we focused on chemicals belonging to two later classes. These are glyphosate, the active ingredient of Roundup herbicide, and glufosinate, the active compound of Basta herbicide. Both chosen herbicides are extensively used worldwide and show some similarities. They are systemic, non-selective, broad spectrum post-emergence herbicides, relatively non-toxic to environment. Moreover, for both herbicides there have been developed herbicide-resistant plants but glyphosate-resistant crops dominate on the world (Service, 2007). Cultivation of herbicide-resistant plants results in the expanded use of these non-selective herbicides. Hence, there is an increasing concern regarding the potential risk of herbicides' pollution and their toxicity to non-target organisms (Blacburn & Boutin, 2003; Boutin et al., 2004; Tesfamariam et al., 2009). The researches concerning the potential phytotoxicity of herbicides to non-target higher plants examine two main aspects of this problem. On the one side, there is investigated the influence of simulated spray drift of herbicides on non-target plants. On the other side, the investigations are conducted to determine the effect of soil applied herbicides (the residues of herbicibes in soil) on plants. In our study we investigated the latter aspect. Herbicides can reach the soil via foliar wash off, undirected spray drift contamination and decomposition of treated plant residues (Tesfamariam et al., 2009). Besides, increased glyphosate concentration in soil may originate from plant roots and exuding process (Laitinen et al., 2007). For our experiment we chose cucumber (*Cucumis sativus* L.), that is very important and popular species in horticultural production. Both herbicides caused visible inhibition of plant growth but there were not observed such injury symptoms as chlorosis or necrosis. Nevertheless, shoots of cucumber exposed to Roundup showed marked symptoms of wilting. Basta and Roundup similarly affected height and fresh weight of hypocotyls and fresh weight of cotyledons. Cucumber roots were very sensitive to Roundup and their fresh weight did not exceed 20% compared to the control plants and was nearly 4-fold lower (on the fresh weight basis) than under Basta treatment. For both herbicides and all organs (roots, hypocotyls, cotyledons) a percentage inhibition in fresh matter accumulation was higher than that in dry matter (Fig. 1A, B, C). This evidently indicates that both Basta and Roundup present in rooting medium markedly disrupt water relations in cucumber tissues. Results in Tab. 1 also show that water status of cucumber organs was strongly impaired, particularly in seedlings exposed to Roundup. These results indicate that the observed inhibition of plant growth could be, in part, caused by the disturbances in water status in cucumber tissues. Negative effect of soil-applied glyphosate and glufosinate on growth of non-target plants was also observed in experiments on sunflower, tomato, maize and cucumber seedlings (Sacała et al., 1999; Sacała et al., 2008;

Tesfamariam et al., 2009; You & Barker, 2005). The most literature data show that application of glufosinate and glyphosate resulted in a decrease in chlorophyll content in plants (Cakmak et al., 2009, Kielak et al., 2011; Pline et al., 1999; Reddy et al., 2000; Zaidi et al., 2005). Our results don't agree with those mentioned above. Cotyledons of cucumber seedlings exposed to herbicides were dark-green and contained more both chlorophyll and carotenoids than non-treated plants (Tab. 2). This indicates that at early stage of cucumber growth (the stadium of fully developped cotyledons) synthesis of chlorophyll is not inhibited by examined herbicides and the ratio of chlorophyll a to chlorophyll b is not impaired. Additionally, increase in total pool of carotenoids in cucumber cotyledons could be an adaptive feature preventing photooxidative damages in chloroplasts. Some researchers express a view that a loss of chlorophyll under stress conditions may be a positive symptom preventing photoinhibition of the photosynthetic apparatus (Maslova & Popova 1993). As mentioned in the Introduction, secondary phytotoxic effect caused by glyphosate is connected with the inhibition of the porphyrin precursor synthesis - δ-aminolevulinic acid – that results in an inhibition in chlorophyll biosynthesis and visible symptoms of chlorosis (Pline et al., 1999). Our results showed that glyphosate applied to the rooting medium did not disturb chlorophyll synthesis in cotyledons but stopped its biosynthesis in leaf. It is worth to note, that the first hardly emerged leaf of cucumber was completely yellow. Moldes et al. (2008) maintain that plants with low constitutive level of chlorophyll might have slower reduction in chlorophyll synthesis rate induced by glyphosate. Whereas, Wong (2002) indicated that the reduction in chlorophyll a content may be dependent on glyphosate concentration and concentration of 2 mg/l or more caused a significant decrease in the level of chlorophyll a.

Both examined herbicides disrupt nitrogen metabolism. The primary disruptions caused by these herbicides take place at different points of the nitrogen assimilation pathway but there may also appear secondary effects connected with the impairment of protein synthesis and functioning of others enzymes involved in nitrate assimilation. Hence, in our study we assumed that nitrate reductase activity and protein concentration will be the common indices of disturbances in nitrogen metabolism caused by both Roundup and Basta. Nitrate reduction in plant tissues is a fundamental process in nitrogen assimilation. The first reaction in nitrate reduction is catalyzed by nitrate reductase. Activity of this enzyme is precisely regulated and controlled by different internal and environmental factors. Among the internal factors closely connected with nitrogen metabolism there are: availability of substrate NO_3^-, concentration of ammonium and the level of end products of nitrate assimilation, mainly glutamine and glutamate (Tischner, 2000). Nitrate stimulates nitrate reductase activity, whereas ammonium and nitrogen compounds accumulated in plants' tissues may repress nitrate assimilation. Reddy et al. (2010) examined the influence of glyphosate simulated drift effect on non-glyphosate-resistant corn. They indicated that glyphosate drift significantly reduced both leaf nitrogen and nitrate reductase activity. Maximum reduction in nitrate reductase activity amounted to 64% compared to non-treated plants. Similar results were obtained by Bellaloui et al. (2006) for non-glyphosate-resistant soybean. Moreover, these researchers indicated that nitrate reductase activity in soybean roots, opposite to the aboveground organs (leaves and stems), was not affected by glyphosate application. Additionally, when glyphosate was applied 6 weeks after planting there was observed large (approximately 2-fold) increase in enzyme activity compared to the untreated control plants. We obtained similar results and we observed nearly 2-fold increase in nitrate reductase activity in cucumber roots growing in medium contained Roundup compared to control plants. In our previous study (Sacała et al., 1999) we also demonstrated that Roundup significantly increased nitrate reductase activity in roots of maize and cucumber var. Wisconsin compared to non-treated plants, whereas roots' growth was strongly suppressed. Basta also caused significant increase in nitrate reductase activity in roots and decrease in cotyledons but observed reduction was not such dramatic as in the case of Roundup application (Tab. 4). Residual nitrate reductase activity in cotyledons of cucumber exposed to glyphosate may result from a loss of enzyme protein (inhibition of enzyme synthesis and inactivation/degradation of already synthesized enzymes), and in part, from the shortage of available reductants for NO_3^- reduction (enzyme activity was assayed at the endogenous level of NADH in plant cells). Under Basta treatment decrease in nitrate reductase activity also could be caused by low availability of NADH. On the other side, it can be assumed that ammonium accumulated in plant cells was important factor repressing this enzyme. Inhibition of glutamine synthetase causes glutamine deficiency and excessive accumulation of ammonium in plant tissues (Pornprom, et al., 2003, Sacala et al., 2008; Wakabayashi & Böger, 2004). All these results show that nitrate reductase from plant roots is less sensitive to examined herbicides than that in cotyledons. Increase in nitrate reductase activity in roots of cucumber exposed to herbicide may be considered as an adaptive feature allowing to compensate a reduction in cotyledons. Bellaloui et al. (2006) also maintain that the assimilation of nitrate in the roots helps minimize the stress effect of herbicide on plant growth. But it is worth to note, that observed increase in enzyme activity, although relatively high, was too low to sustain nitrate reduction in the whole plant at a level similar to the control plants.

Both examined herbicides destroy biosynthesis of intrinsic amino acids and in turn may disrupt synthesis of protein and other N-containing compounds. It is very interesting that in cucumber seedlings, growing in medium contained herbicides, any decrease in protein concentration was not observed (Tab. 3). Basta did not change protein concentration in roots and cotyledons whereas Roundup markedly increased its amount. The largest increase was noticed in roots in the pool of soluble protein. In this case protein concentration was 166% of the value in the control plants. It can be assumed that in this pool are specific proteins defined as stress-associated proteins. Biosynthesis of these proteins may be induced by exposure to stress factors or they may be present constitutively at low level and are elevated in unfavourable conditions. This increased pool of protein might also include the enzymes involved in antioxidant defense mechanisms (Moldes et al., 2008). As mentioned above, Roundup significantly decreased water content in roots and simultaneously increased markedly nitrate reductase activity. Literature data state that nitrate reductase is characterized by high sensitivity to water deficit and its activity rapidly falls under water stress (Barathi, et al., 2001; Burman, et al. 2004). It is possible that some stress proteins protect root enzymes and their mRNA against negative influence of water deficit and directly against damage caused by Roundup. Concentration of insoluble protein also significantly increased although this rise was lower than in the case of soluble protein (16% compared to the control). These changes indicate that cucumber roots activate the biochemical mechanisms that may prevent damage caused by Roundup. It can be assumed that roots being directly exposed to herbicides increase protein synthesis, particularly its metabolically active fraction - soluble protein. However, synthesis of structural (insoluble) protein was also intensified. The response of cotyledons was slight different. Protein concentration in these organs also increased in comparison to non-treated plants but statistically significant increase was noticed in the fraction of insoluble protein and it was similar to that observed in cucumber roots. Proteins accumulated in plant tissues amongst several functions mentioned above may also serve as a storage form of nitrogen that can be utilized within a recovery of growth when stress is over.

Presented results show that relatively low concentrations of Roundup and Basta significantly inhibit growth of cucumber and cause profound changes in protein synthesis and nitrate reductase activity. The changes induced by Basta were lower than those caused by Roundup but concentration of Basta in nutrient solution was 6-fold lower (comparing the molal concentrations) than that of Roundup.

In light of the obtained results, as well as literature data about some weaknesses and risk of herbicide application (Baylis, 2000; Blackburn and Boutin, 2003; Laitinen et al., 2007) it can be concluded, that further increased application of herbicides may have strong stressing effect on non-target plants.

5. Conclusion

Obtained results show that small amount of the available herbicides' residues in root zone may effectively inhibit growth of non-target plants and significantly change their metabolism. Examined plant – cucumber at early phase of growth – is very sensitive to both herbicides but particularly to Roundup. Roundup considerably suppresses the growth of cucumber roots, lowers water content in all organs and dramatically decreases nitrate reductase activity in cotyledons.

The changes in biochemical parameters induced by Basta are smaller than that caused by Roundup.

6. References

Barathi, P., Sundar, D., Ramachandra, A. & Reddy A. (2001). Changes in mulberry leaf metabolism in response to water stress. *Biologia Plantarum*, Vol. 44, No 1, pp. 83-87, ISSN 0006-3134

Baylis, A. (2000). Why glyphosate is a global herbicide: strengths, weaknesses and prospects. *Pest Management Science*, Vol. 56, No 4, (April 2000), pp. 299-308, ISSN 1526-498X

Bellaloui, N.; Reddy, K.N.; Zablotowicz, R.M. & Mengistu, A. (2006). Simulated glyphosate drift influences nitrate assimilation and nitrogen fixation in non-glyphosate-resistant soybean. *Journal of Agricultural and Food Chemistry*, Vol. 54, No 9, (May 2006), pp. 3357-3364, ISSN 0021-8561

Blackburn, L.G. & Boutin, C. (2003). Subtle effects of herbicide use in the context of genetically modified crops: A case study with Glyphosate (Roundup®). *Ecotoxicology*, Vol. 12, No 1-4 (February 2003), pp. 271-285, ISSN 0963-9292

Boutin, C., Elmegaard, N. & Kler, C. (2004). Toxicity testing of fifteen non-crop plant species with six herbicides in greenhouse experiment: Implications for risk assessment. *Ecotoxicology*, Vol. 13, No 4 (May 2004), pp. 349-369, ISSN 0963-9292

Burman, U., Garg, B.K. & Kathju, S. (2004). Interactive effecs of thiourea and phosphorus on clusterbean under water stress. *Biologia Plantarum*, Vol. 48, No 1, pp. 61-65, ISSN 0006-3134

Cakmak, I., Yazici, A., Tutus Y. & Ozturk, L. (2009). Glyphosate reduced seed and leaf concentrations of calcium, manganese, magnesium, and iron in non-glyphosate resistant soybean. *European Journal of Agronomy*, Vol. 31, No 3 (October 2009), pp. 114-119, ISSN 1161-0301

Cobb, A. (1992). The inhibition of amino acid biosynthesis, In: *Herbicides and Plant Physiology*, Chapman & Hall, pp. 126-144, ISBN 0 412 43860 7, Great Britain

Hanke, I.; Wittmer, I., Bischofberger S., Stamm, Ch. & Singer, H. (2010). Relevance of urban glyphosate use for surface water quality. *Chemosphere*, Vol. 81, No 3 (September 2010), pp. 422-429, ISSN 0045-6535

Jaworski, E.G. (1971). Nitrate reductase assay in intact plant tissues. *Biochemical and Biophysical Research Communications*, Vol. 43, No 6, (June 1971), pp. 1274-1279, (April 2000), ISSN 0006-291X

Kaiser, W.M. & Huber, S.C. (1994). Post-translational regulation of nitrate reductase in higher plants. *Plant Physiology*, Vol. 106, No 3 (November 1994), pp. 817-821, ISSN 0032-0889

Kielak,, E., Sempruch C., Mioduszewska, H., Klocek ,J. & Leszczyński, B. (2011). Phytotoxicity of Roundup Ultra 360 SL in aquatic ecosystems: Biochemical evaluation with duckweed (*Lemna minor* L.) as a model plant. *Pesticide Biochemistry and Physiology* Vol. 99, No 3 (March 2011), pp. 237-243, ISSN 0048-3575

Laitinen, P.; Rämö, S. & Siimes, K. (2007). Glyphosate translocation from plants to soil – does this constitute a significant proportion of residues in soil? *Plant and Soil*, Vol. 300, No 1-2 (November 2007), pp. 51-60, ISSN 0032-079X

Lichtenthaler, H.K. (1987). Chlorophylls and carotenoids: Pigments of photosynthetic biomembranes. *Methods of Enzymology, Plant Cell Membranes*, Packer & Douce, 148, pp. 350-382, ISBN 978-0-12-182048-0

Lowry, O.H.; Rosenbrough, N.J.; Farr, A.L. & Randall R.J. (1951). Protein measurement with Folin-phenol reagent. *Journal of Biological Chemistry*, Vol. 193, No. 1, (November 1951), pp. 265-275, ISSN 0021-9258

Maslova, T.G. & Popova, I.A. (1993). Adaptive properties of the plant pigment systems. *Photosynthetica*, Vol. 29, pp. 195-203, ISSN 0300-3604

Moldes, C.A. Medici, L.O., Abrahäo, O.S., Tsai, S.M. & Azevedo R.A. (2008). Biochemical responses of glyphosate resistant and susceptible soybean plants exposed to glyphosate. *Acta Physiologia Plantarum*, Vol. 30, No 4 (July 2008), pp. 469-479, ISSN 0137-5881

Pline, W.A., Wu J. & Hatzios, K.K. (1999). Effects of temperature and chemical additives on the response of transgenic herbicide-resistant soybeans to glufosinate and glyphosate applications. *Pesticide Biochemistry and Physiology*, Vol. 65, No 2 (October 1999), pp. 119-131, ISSN 0048-3575

Pornprom, T., Champoo, J. & Grace, B. (2003). Glufosinate tolerance in hybrid corn varieties based on decreasing ammonia accumulation. *Weed biology and Management*, Vol. 3, No 1 (March 2003), pp. 41-45, ISSN 1444-6162

Reddy, K.N., Bellaloui, N. & Zablotowicz, R.M. (2010). Glyphosate effect on shikimate, nitrate reductase activity, yield and seed composition in corn. *Journal of Agricultural and Food Chemistry*, Vol. 58, No 6, (March 2010), pp. 3646-3650, ISSN 0021-8561

Sacała, E., Demczuk, A. & Grzyś, E. (1999). Phytotoxicity of Roundup herbicide under saline environment conditions (in Polish, with English abstract). *Zeszyty Problemowe Postępów Nauk Rolniczych* 469, pp. 473-480, ISSN 0084-5477

Sacała, E., Podgórska-Lesiak, M. & Demczuk, A. (2008). Glufosinate phytotoxicity to maize under salt stress conditions. *Polish Journal of Environmental Studies*, Vol. 17, No 6, (2008), pp. 993-996, ISSN 1230-1485

Service, R.E. (2007). A growing threat down on the farm. *Science*, Vol. 316 (May 2007), pp. 1114-1117, ISSN 0036-8075

Sitt, M.; Müller, C.; Matt, P.; Gibon, Y.; Carillo, P.; Morcuende, R.; Scheible, W.-R. & Krapp, A. (2002). Step towards an integrated view of nitrogen metabolism. *Journal of Experimental Botany*, Vol. 53, No 370 (April 2002), pp. 959-970, ISSN 0022-0957

Tan, S., Evans, R. & Singh, B. (2006). Herbicidal inhibitors of amino acid biosynthesis and herbicide-tolerant crops. *Amino Acids*, Vol. 30, No 2 (March 2006), pp. 195-204, ISSN 0939-4451

Tesfamariam, T., Bott, S., Cakmak,, I., Römheld V. & Neumann, G. (2009). Glyphosate in the rhizosphere – Role of waiting times and different glyphosate binding forms in soils for phytotoxicity to non-target plants. *European Journal of Agronomy*, Vol. 31, No 3 (October 2009), pp. 126-132, ISSN 1161-0301

Tischner, R. (2000). Nitrate uptake and reduction in higher and lower plants. *Plant, Cell and Environment*, Vol. 23, No 10 (October 2000), pp.1005-1024, ISSN 0140-7791

Wakabayashi, K. & Böger, P. (2004). Phytotoxic sites of action for molecular design of modern herbicides (part 2): Amino acid, lipid and cell wall biosynthesis, and other targets for future herbicides. *Weed Biology and Management,* Vol. 4, No 2 (June 2004), pp. 59-70, ISSN 144-6162

Wong, P.K. (2000). Effects of 2,4-D glyphosate and paraquat on growth, photosynthesis and chlorophyll-a synthesis of *Scenedesmus quadricauda* Berb 614. *Chemosphere*, Vol. 41, No 1-2 (July 2000), pp. 177-182, ISSN 0045-6535

Woodburn, A.T. (2000). Glyphosate: production, pricing and use worldwide. *Pest Management Science*, Vol. 56, No 4 (April 2000), pp. 309-312, ISSN 1526-498X

Vereecken , H. 2005. Mobility and leaching of glyphosate: a review. *Pest Management Science*, Vol. 63, No 12 (December 2005), pp.1139-1151, ISSN 1526-498X

You, W. & Barker, A.V. (2005). Effects of soil-applied glufosinate-ammonium on tomato plant growth and ammonium accumulation *Communications in Soil Science and Plant Analysis*, Vol. 35, No 13 & 14 (January 2005), pp. 1945-1955, ISSN 0010-3624

Zaidi, A., Khan, M.S. & Rizvi, P.Q. (2005). Effect of herbicides on growth, nodulation and nitrogen content in greengram. *Agronomy for Sustainable Development*, Vol. 25, No 4 (October-December 2005), pp. 497-504, ISSN 1774-0746

Immunosensors Based on Interdigitated Electrodes for the Detection and Quantification of Pesticides in Food

E. Valera[1,2] and A Rodríguez[3]
[1]Applied Molecular Receptors Group (AMRg), IQAC-CSIC, Barcelona,
[2]CIBER de Bioingeniería, Biomateriales y Nanomedicina (CIBER-BBN), Barcelona,
[3]Micro and Nano Technologies Group (MNTg), Departament d'Enginyeria Electrònica,
Universitat Politècnica de Catalunya, C/., Barcelona,
Spain

1. Introduction

The use of substances addressed to prevent, destroy or control pests has helped to protect mankind against many different types of pests. Pesticides have been used since ancient times but the discovery of new chemicals boosted their use in the second half of the 20th century. Pesticides used in agriculture made possible to hugely increase and improve production, helping to control insects, bacteria, fungi, herbs, etc. The benefits of their use and their impact in the economy were great and, therefore, the use of pesticides spread rapidly all around the world. The intensive use of pesticides raised concerns about their possible negative effects. Thus, extensive research has been carried out on the effect of pesticides on health, environmental pollution and impact on wildlife. This fact has leaded to the development of new international and national regulations for the rational use of pesticides.

In relation to health, pesticides can penetrate into human bodies in different ways: they can be inhaled by breathing, they can enter through the skin or wounds, and obviously they can be ingested by eating foods containing residual amounts of pesticides. Pesticides are not necessarily poisonous but they may be toxic. The effect of each pesticide on human health depends on the dose and time of contact. Regulations specify maximum residue levels, highest concentration of each pesticide that is allowed to be present in foods. Thus, the European Community has established maximum residue levels for the different pesticides (Council Directives 76/895/EEC, 86/362/EEC, 86/363/EEC and 90/642/EEC) and particularly for atrazine (Directive 93/58/EEC), in various foodstuff products.

Many types of sensors have been developed for the detection and quantification of pesticides and traces of them. The MRL of a given pesticide, which often lies in the order of few tens of ppbs, determines the minimum sensitivity of these sensors. Also the quantification, or the detection, of one substance in the complex chemical matrix of some foods as wine, milk or juices also poses important requirements to their selectivity. Traditionally, samples of the products to be analyzed were taken to laboratories where precise apparatus based on chromatography methods such as HPLC or GC/MS are able to perform the analysis. These processes of analysis involving the transport to the samples, the analysis and the communication of results, can be time consuming and may be expensive.

It is of high interest to analyze food products at different points of the food chain: recollection, transport, storage, consumption, etc. In particular, with foods like wine, milk, juices, etc, producers take their products to common collection points, where they are mixed with products of other producers. Analysis should be done at this point in order to reject contaminated products before mixing them with good ones.

In order to be able to fabricate these sensors, several conditions need to be accomplished: (i) to maintain the required selectivity and sensitivity, (ii) sensors need to be fast and (iii) the price of each analysis has to be acceptable. Besides, it would be interesting that the sensing system could be portable or at least, compact, and also that the process of analysis would not require specialized personnel, providing a result of simple interpretation with a minimum of sample manipulation. The previous statements are based on a careful market study of the wine industry performed during the initial stage of the European Project GoodFood (FP6-IST-1-508744-IP).

The two types of immunosensors studied in this chapter have been oriented towards the detection of small amount of pesticides residues in wine samples. They are based on the use of: i) interdigitated μ-electrodes (IDμE's) arrays; and ii) bioreagents specifically developed (antigen, antibody).

The main characterization method used in the study of these immunosensors has been impedance spectroscopy in a wide range of frequencies (40Hz – 1MHz). Nevertheless, besides impedance spectroscopy, the immunosensors developed have been also characterized by means of other impedance methods as well as chemical affinity methods in order to contrast their performances. The immunosensors developed have been named:

i.   Impedimetric immunosensor;
ii. Conductimetric immunosensor.

The nomenclature used is related to the detection methods applied in the present work. In the case of the impedimetric immunosensor the detection method is based on impedimetric measurements (in a wide range of frequencies), whereas in the case of the conductimetric immunosensor, the detection method is based on conductimetric measurements (DC measurements). For the case of the conductimetric immunosensor, conductimetric measurements as detection method are possible because this sensor is labelled with gold nanoparticles.

2. Description of the immunosensors

In this section, the basic ideas underlying the structure, functionalization and working mechanisms of the biosensors treated in this chapter are described.

2.1 Interdigitated μ-electrodes (IDμE's)

As it has been commented before, interdigitated μ-electrodes (IDμE's) were used as transducers for the immunosensors presented in this chapter. Interdigitated μ-electrodes are two coplanar electrodes (that works as counter and working electrodes) which have equal surface areas and each is presumed to contribute equally to the measured network impedance. The procedure of the electrodes fabrication is as follows:

Thin Au/Cr (~200 nm thickness) interdigitated μ-electrodes (IDμE's) with, 3.85 μm thick with electrode gap of 6.8 μm were patterned on a Pyrex 7740 glass substrate (purchased from Pröazisions Glas & Optik GmbH, 0.7mm (±0.05) thickness). The chromium layer is much thinner than the gold layer and it is deposited prior to gold just to improve the adhesion of the gold to the Pyrex substrate. Before metal deposition, the Pyrex substrate was cleaned using absolute ethanol. The metal deposition was performed by means of sputtering deposition and the interdigitated μ-electrodes were then patterned on the Pyrex substrate by a photolithographic metal etch process. For the immunosensor measurements, arrays consisting on six IDμE's organized on a 0.99 cm^2 area were constructed.

Before functionalization, the samples were first cleaned in a solution of ethanol absolute 70% and Milli-Q water 30%. Then, the samples were plunged for 12 h in a solution of NaOH 2.5% in Milli-Q water. Afterwards, the 12 h the samples were rinsed in 100mL of Milli-Q water in order to neutralize the action of the NaOH.

Finally, the arrays of IDμE's were dried with ethanol and N$_2$.

2.2 Impedimetric immunosensor

The impedimetric immunosensor is a robust and **label-free device** based on the use IDμE's arrays, bioreagents specifically developed and on the **impedimetric change** that occurs when the immunoassay is performed on the electrodes surface.

The assay of detection relies on the immunochemical competitive reaction between the pesticide residues and the immobilized antigen on IDμE's for a small amount of the specific antibody. The detection of a small number of molecules of pesticide residues is performed under competitive conditions involving the competition between the free pesticide (analyte) and a fixed amount of coated antigen for a limited amount (low concentration) of antibody (Ab). At the end of the reaction, the amount of Ab captured on the IDμE surface and hence the free antigen (analyte), is determined.

This competitive assay is fundamental in the immunosensor concept, because, as it can be clearly seen in Figure 1, the immunosensor actually does not measure an amount of pesticide; instead it measures an amount of antibody (related to the target pesticide). Thus, the change in the impedance is due to the addition of antibody in the sensor surface, and not to the addition of molecules of pesticides. This approach has an important effect on the sensitivity of these immunosensors, because the molecules of antibody are much bigger than the molecules of pesticide and their effect on the impedance of the device is much higher. This feature represents an important advantage in comparison with other impedimetric immunosensors reported previously [1-3]. As a consequence, authors of these works must reduce the electrode size to nanometer scale [1], or otherwise their limits of detection can only achieve tens of ppbs [2, 3].

Immunosensor functionalization consists on two main steps: i) the coating antigen (CA) immobilization; and ii) the specific antibody capture. These steps will be schematically shown below. The addition of pesticide in residual concentrations, during the antibody capture step, makes that a fraction of initial antibodies will be evacuated from the device (Figure 1). Thus, the change in the antibody concentration is equivalent to the pesticide concentration used.

As it is shown herein below, this immunosensor is sensitive to the chemical changes produced at the surface of its interdigitated μ-electrodes, and hence the impedance measured will change following the changes of: (i) the concentration of the immobilized antigen, (ii) the amount of the captured antibody and (iii) the competitive equilibrium between analyte, specific antibody and the competitor antigen.

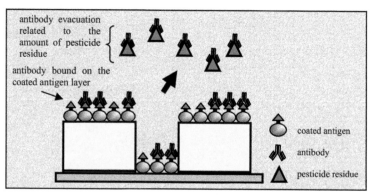

Fig. 1. Immunosensor reaction. An amount of the specific antibody is bound on the coated antigen layer. Other quantity is evacuated of the IDμE's; this amount is related to the pesticide concentration.

2.3 Conductimetric immunosensor

The conductimetric immunosensor is a **labelled device**. This device is also based on the use IDμE's arrays, bioreagents specifically developed but, in addition, it includes a **secondary antibodies labelled with gold nanoparticles**. In consequence, in this case the detection principle is based on a **conductimetric change** which occurs when the secondary antibody is deposited on the electrodes surface, after the immunoassay.

As in the previous case, the assay of detection also relies on the immunochemical competitive reaction between the pesticide residues and the immobilized antigen on IDμE's for a small amount of the specific antibody. However in this case, a secondary antibody (Ab_2) is included (Figure 2). These secondary antibodies, linked to the gold particles, constitute a conductive film between the electrodes. Thus, the conductance of this film will depend on the concentration of gold labelled antibodies.

The functionalization of this immunosensor consists in three main steps: i) the coating antigen (CA) immobilization; ii) the specific antibody capture (Ab_1); and iii) the capture of a non- specific antibody (Ab_2) labelled with gold nanoparticles.

The detection of free pesticide still depends on the competition between the analyte and a fixed amount of CA for a low concentration of Ab_1. After that, Ab_2 is included and linked to Ab_1, then the Ab_2 concentration (and, as a consequence, the amount of gold particles) is related to the Ab_1 concentration included. Therefore, the concentration of the free pesticide tested is related to the amount of gold nanoparticles. Again, the immunosensor does not measure directly the quantity of pesticide; in this case an amount of gold particles related to the amount of pesticide, is measured. This procedure is schematically shown in Figure 3.

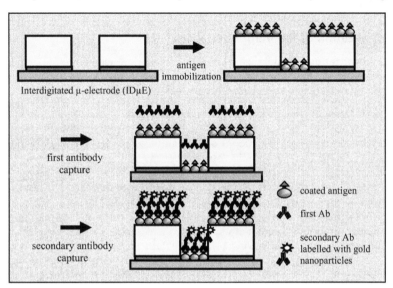

Fig. 2. Schematic diagram of the complete assay system performed on the IDµE's for the conductimetric immunosensor.

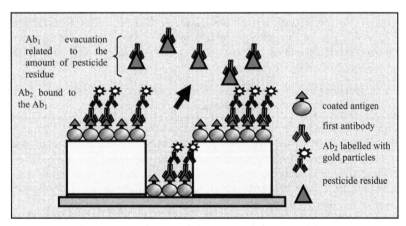

Fig. 3. Immunosensor reaction. An amount of the secondary antibody (Ab$_2$) is bound to the specific antibody (Ab$_1$). Previously, an amount of Ab$_1$ (related to the pesticide residue concentration) was evacuated of the IDµE's; the amount of gold nanoparticles is related to the pesticide residue concentration.

Comparing the functionalization procedures of both immunosensors, apparently the only difference is the inclusion of the gold nanoparticles. However, the consequence of this difference is not only related to the detection method. It is related to the fact that the inclusion of the gold nanoparticles causes a very different distribution of electric field in comparison to the case of having only the fingers of the interdigitated electrodes. In this new structure, gold particles act as new *small fingers*, reducing the gap of the interdigitated µ-electrodes [4].

3. Functionalization of the Immunosensors

The biofunctionalization (immobilization) of the biological element onto the transducer surface is required for the immunosensors development. In this section, the functionalization of the immunosensors is explained for all the cases proposed.

Sensor solid surfaces are in general solid inorganic materials not suitable for immobilizing biomolecules. Hence, further modification is required to adapt them for the immobilization of biomolecules. In addition, functional sensor surfaces place several demands such as biocompatibility, homogeneity, stability; specificity; and functionality. Thus, a challenge in biosensor development is to construct adequate surfaces as well as to design molecules suitable for site-directed immobilization. Surface architecture depends on the nature of the transducer and on the features of the biomolecule, as well as the type of measurements to be done [5-8]. The surface has to be activated appropriately for further tethering of the proteins with a particular immobilization method. Subsequent layers can be generated in place, textured following specific demands. A key problem is the non-specific attachment of molecules, sometimes present in the matrix where measurements need to be made, to the surface of the sensor. This happens on any kind of surface, but particularly, gold is very well suited to capture non-specifically organic molecules and components from the media. For that, the affinity of the antibodies as well as the adequate functionalization of the surface (electrodes and gap) is very important.

Immunosensors functionalization is based on the capture of antibodies specifically developed. The function of the antibody is the capture of the antigen and to form with it a complex antibody with the aim to exclude intruders. In addition to the antibody, the immunosensor reaction implies the presence of a coated antigen and the analyte.

The dielectric properties of the biological systems are very remarkable. Thanks to this important characteristic, these devices can exhibit a good impedimetric response.

The detection of pesticides in very low concentrations relies in a competitive reaction between the analyte and an immobilized protein (coated antigen) supported analog. Over the coated layer of antigen, the free specific antibody is captured by affinity. In the case of the conductimetric immunosensor, a secondary antibody labelled with a gold nanoparticle is attached to the primary specific antibody in order to amplify the affinity event and obtain a good conductive response.

In the case of both immunosensors described in this chapter, the method of immobilization used is covalent immobilization, and the procedure is explained hereinafter.

3.1 Functionalization of the impedimetric immunosensor

In the case of the impedimetric immunosensor, the chemical changes on the sensor surface follow four steps, two previous steps for the surface functionalization and two more for the immunosensor reaction:

i.   Step I, protection of interdigitated µ-electrodes with N-acetylcysteamine;
ii.   Step II, immunosensor surface functionalization with GPTS;
iii.   Step III, covalent immobilization of the antigen on the IDµE;
iv.   Step IV, specific primary antibody (Ab₁) capture in the competition step;

One of the main consequences of the use of covalent immobilization is that the chemical recognition layer is only deposited on the gap of the interdigitated µ-electrodes, because it is the substrate surface (and not the electrodes) which is functionalized (Step II). The immunosensor surface functionalization is shown in Figure 4, whereas the complete chemical procedures are schematically described in Figure 5.

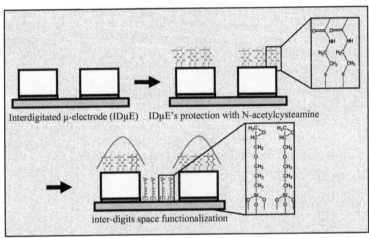

Fig. 4. Schematic diagram of: i) protection of interdigitated µ-electrodes with N-acetylcysteamine; and ii) immunosensor surface functionalization with GPTS.

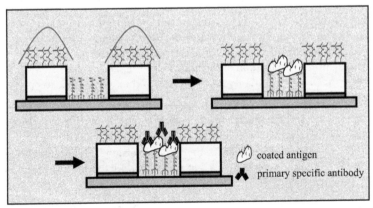

Fig. 5. Schematic diagram of the complete assay system performed on the IDµE's for the impedimetric immunosensor.

Activation of gold surfaces is readily and specifically performed using thiol-chemistry. N-acetylcysteamine is used to cover the gold electrodes and to protect the sensor from undesired non-specific absorptions (Step I). The resulting Au-S bond grants good stability of the deposited surface layer. In this case, the surface texture of the IDµE defines the template for deposition of layers, since the gold fingers have been deposited on a solid support such as glass with the necessary controlled geometry. This is not the case for glass material that serves as support. Silane-chemistry is the most used activation procedure to functionalize the surface for subsequent covalent coupling of the biomolecules. Thus, in a next step (Step II) the PYREX substrate is derivatized with 3-glycidoxypropyl trimethoxysilane (GPTS). The epoxy group provides the necessary reactivity for further attachment of the bioreagents through a nuchelophylic attack of functional groups of the biomolecule such as the amino groups of the lysine residues. As it has been reported [9], the concentration of the silane (2.5% in anhydrous ethanol) and the reaction time (12 h) are important for the formation of a homogenous and molecularly smooth epoxysilane layer on the PYREX substrate. Finally, covalent immobilization of the pesticide antigen 2d-BSA is performed on the surface of the interdigitated µ-electrodes via the amino groups of the lysine residues by reaction with the epoxy groups of the surface (Step III).

The impedance of the microelectrodes after the functionalization steps previous to the Step IV will constitute a reference value, and in what follows will be denoted as blank. Blank implies the substrate impedance (Z_{board}); the ohmic resistance of the electrolyte (Rs); the contribution to the impedance of the N-acetylcysteamine (Z_N); the contribution of the GPTS (Z_{GPTS}); and the impedance of the antigen (Z_{AT}). Therefore, the impedance of interest is the increment of impedance between Step III and Step IV ($Z_{Ab} - [Z_N + Z_{GPTS} + Z_{AT}]$).

3.2 Functionalization of the conductimetric immunosensor

As in the case of the impedimetric immunosensor, a covalent immobilization technique was also applied to the conductimetric immunosensor. The chemical changes on the conductimetric immunosensor surface follow five steps, two previous steps for the immunosensor surface functionalization and other three steps for the immunosensor reaction:

i.   Step I, protection of interdigitated µ-electrodes with N-acetylcysteamine;
ii. Step II, immunosensor surface functionalization with GPTS;
iii. Step III, covalent immobilization of the antigen on the IDµE;
iv. Step IV, specific primary antibody (Ab$_1$) capture in the competition step;
v.   Step V, secondary labelled with gold antibody (Ab$_2$) capture.

As in the case of the impedimetric immunosensor, the chemical recognition layer was deposited only on the gap of the interdigitated µ-electrodes. The immunosensor surface functionalization was the same applied to the impedimetric immunosensor (Figure 4). The complete functionalization procedures of the conductimetric immunosensor are schematically shown in Figure 6.

As in the case of the impedimetric immunosensor, the blank was also defined. In this case, blank is the contribution in impedance related to the functionalization steps previous to the Step V. Thus, blank implies the substrate impedance (Z_{board}); the ohmic resistance of the electrolyte (Rs); the contribution to the impedance of the N-acetylcysteamine (Z_N); the contribution of the GPTS (Z_{GPTS}); the impedance of the antigen (Z_{AT}); and the impedance of the primary antibody (Z_{Ab1}). Therefore, the impedance of interest is the delta (variation) of impedance between Step IV and Step V ($[Z_{Ab2 + gold\ particles}] - [Z_{Ab} + Z_N + Z_{GPTS} + Z_{AT}]$).

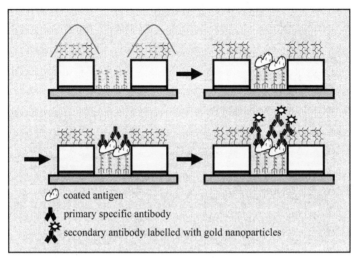

Fig. 6. Schematic diagram of the complete assay system performed on the IDµE's for the conductimetric immunosensor.

4. Pesticide residues detection in buffer samples

Both types of immunosensors described in previous sections (impedimetric and conductimetric) have been tested for the detection of free pesticide in buffer samples. For that, **atrazine**, a widely used pesticide in the wine industry, as well as for the test of novel biosensors [10-14], has been used as pesticide of test.

Atrazine and related triazines such as ametryn, propazine, prometryn, prometon, simazine and terbutryn, are widely used selective herbicides for the control of annual grasses and broadleaved weeds. Therefore, herbicide residues can contaminate crops, wells, and streams due to spills, spraying and run-off. They have often been found in drinking water, and therefore, they are a potential threat for the public health [15-18]. The European Community has established maximum residue level for residues of this herbicide in wine grapes in 50 µg L^{-1}.

The competitive reaction carried out on the interdigitated µ-electrodes has been performed in buffer solution (assay buffer). The performance details of both types of devices as atrazine detectors are detailed below.

4.1 Impedimetric immunosensor

4.1.1 Detection method applied: Impedance spectroscopy

One of the quantitative tools adequate to provide sensitivity graphs is the impedance characterization measurement in a wide frequency range and the fitting of the Nyquist plots of impedance spectra to an equivalent circuit. The use of this technique and Nyquist plots are a very common strategy in the literature for the biosensors characterization [2, 3, 19-23]. By this technique, the concentration of free pesticide should finally be related to the values of at least some of the parameters of the equivalent circuit.

The equivalent circuit used is shown in Figure 7. The circuit is an equivalent electrical circuit that consists of:

i.   contact resistance, Rc;
ii.   capacitance of the IDμE, $C_{IDμE}$;
iii.   the Warburg impedance, Zw;
iv.   the double layer capacitance, Cdl; and
v.   Polarization resistance; Rp.

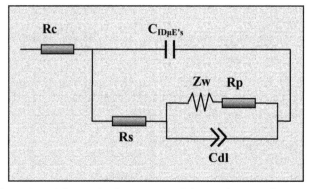

Fig. 7. Simplified version of the equivalent circuit of the implemented system.

Impedance spectroscopy requires a wide bandwidth of work, the application of a periodic small amplitude AC signal, as well as a solution as medium (typically a buffer) for charge transport. For the impedimetric immunosensor characterization, frequencies in the range of 40 Hz – 1 MHz; 0V of polarization potential; a modulation voltage of 25mV amplitude; and a phosphate buffer solution (PBS @ 1.6 μS cm^{-1}) as medium were used.

After performing the measurements, response curves are fitted to the following four-parameter equation [24]:

$$Y = \{\frac{A - B}{1 + (x / C)^{D}}\} + B \tag{1}$$

where A is the absolute maximal signal (zero analyte concentration), B is the absolute minimum signal (infinite analyte concentration), C is the concentration producing 50% of the maximal signal change, and D is the slope at the inflection point of the sigmoid curve.

As an example, the Nyquist plots of impedance spectra of layer-by-layer, under covalent immobilization conditions, are shown in Figure 8. As it can be seen, an excellent fitting between the simulated and experimental spectra has been obtained for all the curves. Curve 8a is the impedance spectrum after 1 μg mL^{-1} concentration of antigen immobilization, while the curve 8b is the impedance spectrum after the specific antibody capture in 1 μg mL^{-1} concentration.

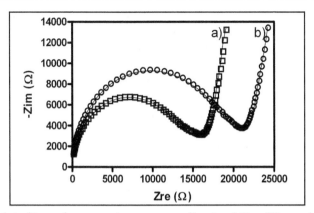

Fig. 8. Nyquist plot of impedance spectra corresponding to: a) Step III: covalent immobilization of the antigen on the IDμE, and b) Step IV: specific antibody (Ab) capture in the competition step, taken in diluted PBS solution without redox couple. Symbols represent the experimental data. Solid curves represent the computer fitting data with the parameters calculated by a commercially available software Zplot/Zview (Scibner Associates Inc.) using the equivalent circuit shown in Figure 7.

4.1.2 Atrazine detection in buffer samples

In order to detect atrazine in buffer samples using the impedimetric immunosensor, buffer samples were doped with different atrazine concentracion in the range of 0.32 – 2000 μg L^{-1}. The experiments were conducted including these concentrations of atrazine during the competition step using different ID's samples for every concentration. The response curve obtained, using the tool of detection in a range of frequencies, the covalent immobilization technique, as well as assay buffer, can be seen in Figure 9.

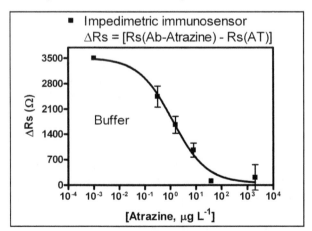

Fig. 9. Response curve of the impedimetric immunosensor, using the covalent immobilization technique, for the atrazine detection in relation with the Rs variation. Buffer solution was used for the competitive reaction. Measures were taken in diluted PBS solution. See Table 1 for the features of the atrazine assay.

All measurements were differential in order to suppress the non-ideal effects related to the geometry or technology of the IDµE's. Likewise, the variation of Rs in relation with atrazine concentration was selected and analyzed from the computer fitting data. In all cases, the change in the value of Rs (denoted ΔRs in Figure 9) was taken between the Step III and Step IV of the impedimetric immunosensor functionalization. Due to the competitive assay performed, ΔRs decreases as the concentration of atrazine increases.

The limit of detection obtained from the response curve shown in Figure 9 was **0.04 µg L^{-1}**. This value is not only far below the MRL required by EC, but it also shows a huge improvement on the results obtained when the passive adsorption was applied.

Features of the atrazine assay	Impedimetric immunosensor (Buffer)
Signal$_{min}$	56.3[b]
Signal$_{max}$	3510[b]
Slope	-0.66
IC$_{50}$, µg L^{-1}	1.23
LOD, µg L^{-1}	**0.04**
R^2	0.91

[a] The parameters are extracted from the four-parameter equation used to fit the standard curve.
[b] Values presented are in Ω.

Table 1. Features of the atrazine assays[a].

4.2 Conductimetric immunosensor

4.2.1 Detection method applied: DC measurements

Although a priori DC measurements do not seem a great contribution as detection method, they represent a very important approach of the conductimetric immunosensor, because these DC measurements are based on the gold nanoparticles attached to the immunosensor. With this method, it was proven that pesticides in residual amounts can be detected by means of simple and inexpensive DC measurements provided gold nanoparticles are included as labels in the immunosensor [25, 26].

When a DC voltage is applied to an interdigitated µ-electrode without gold nanobeads, the DC current obtained is low because the dielectric properties of the bioreagents and the gap that separates the electrodes. Nevertheless, when the same DC voltage is applied to an interdigitated µ-electrode with gold particles attached to it, the DC current which passed through the electrodes grows clearly. Furthermore, if the concentration of gold particles is in relation to the concentration of pesticide, then the current values obtained will also be.

Thus, the measurement of the conductance (DC current) after the inclusion of the gold nanoparticles on the immunosensor is a quantitative tool adequate to provide sensitivity graphs. Then, the concentration of pesticide should finally be related to the amount of gold nanoparticles present on the immunosensor.

On the other hand, detection techniques based on impedimetric measurements were discarded in the case of the conductimetric immunosensor because the inclusion of the gold particles is a handicap for the impedimetric response. An important advantage of the *competitive* assay used is that the impedimetric variation is related to the molecules of antibody, instead the molecules of pesticide (smaller than the antibody molecules). Therefore, applying impedimetric measurements to the conductimetric immunosensors, this advantage is reduced, because the contribution of the gold particles (conductive elements) *subtracts* the secondary antibody contribution.

4.2.2 Atrazine detection in buffer samples

As in the previous case, experiments were carried out including atrazine concentrations between 0.32 to 2000 μg L⁻¹ during the competition step (Step IV) and using different ID's arrays for every concentration. The electrodes were covered by a diluted PBS solution and the measurements were executed to +25 and +100 mVdc bias. DC voltages were chosen under 100 mV bias in order to avoid the electrolysis of water. Likewise, the *blank* was measured and reduced from the data measured after Step V in order to take into account only the contribution of the gold nanoparticles. The results obtained by this method are represented in Figure 10. Again, the atrazine response of the sensor follows an inverse law and hence the response is larger at low concentrations of atrazine. The limits of detection obtained for the atrazine residues detection using the conductimetric immunosensors, when the competitive assay was performed in buffer solution, were **0.446 μg L⁻¹** (100 mVdc bias) and **1.217 μg L⁻¹** (25 mVdc bias), both far below the MRL.

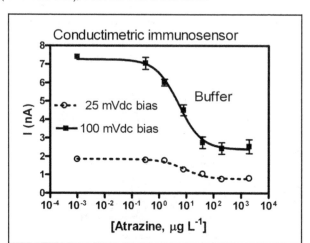

Fig. 10. Response curve of the conductimetric immunosensor, using the covalent immobilization technique, for the atrazine detection in relation with the presence of gold particles (40 nm). Buffer solution was used for the competitive reaction. Measures were taken in diluted PBS solution and the *blank* was reduced. See Table 2 for the features of the atrazine assay.

Features of the atrazine assays[a]	Conductimetric immunosensor (buffer)	
	25 mV	100 mV
IC_{50}, µg L^{-1}	8.47±0.19	5.29±0.14
LOD, µg L^{-1}	**1.217**	**0.466**
R^2	0.89	0.91

[a] The parameters are extracted from the four-parameter equation used to fit the standard curve.

Table 2. Features of the atrazine assays[a].

4.3 New approach: a flexible device

As it was proven in the previous sections, both types of immunosensors described in this chapter have been able to detect residues of atrazine when buffer samples were used for the competitive assay. In both cases, the transducer was supported by a PYREX substrate.

In this section a new approach is introduced, the possibility of low cost, flexible plastic substrates.

4.3.1 Flexible interdigitated µ-electrode (FIDµE)

PYREX properly complies with the conditions of isolation and compatibility necessary, and, furthermore, it gives a great rigidity and stability to the device. Nevertheless, the possibility of a flexible sensor has also been explored. Therefore, the first step to carry out this idea is the development of IDµE's with flexible features. Flexible interdigitated µ-electrodes (FIDµE) for biosensor applications have also been fabricated. A sample of these FIDµE's can be seen in Figure 11.

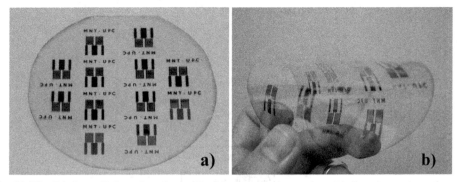

Fig. 11. Flexible interdigitated µ-electrodes fabricated: a) top view; b) demonstration of flexibility.

The flexibility of the FIDµE's is related to the plastic substrate where the electrodes are deposited. The plastic chosen as new substrate was polyethylene naphthalate (PEN), 0.075 mm, purchased from Goodfellow Cambridge Limited. PEN was chosen instead of other plastics such as Polyethylene terephthalate (PET) or Polycarbonate (PC), because it brings together important features that make it compatible with the microelectronic technology such as:

i.   chemical resistance: acids – dilute (good), alcohols (good);
ii. electrical properties: dielectric constant @ 1 MHz (3.2 @ 10 kHz), surface resistivity (10^{14} Ω/sq);
iii. thermal properties: upper working temperature (155 °C).

The fabrication procedure of the FIDµE's is as follows: Thin Au (150 - 200 nm thickness) interdigitated µ-electrodes (IDµE's) with, 30 µm thick with electrode gap of 30 µm were patterned on a PEN substrate. In this case, the chromium layer is avoided because the good adhesion between gold and the PEN substrate. Before metal deposition, the PEN substrate was cleaned using absolute ethanol. The metal deposition was performed by means of sputtering deposition and the interdigitated µ-electrodes were then patterned on the PEN substrate by a photolithographic metal etch process.

FIDµE's solves the problem of the flexible electrodes; nevertheless, in order to obtain an immunosensor, the sensor surface must offer good biochemical behaviour. A material as polyethylene naphthalate, used for electronic applications, is not biocompatible. Therefore, the functionalization by means of covalent immobilization techniques could not be performed on PEN. To solve this problem, a biocompatible layer must be deposited, at least, on the gap surface. For this reason, SiO_2 was deposited by sputtering on the electrodes surface, because silicon oxide surface contains reactive SiOH groups, which can be used for covalent attachment of organic molecules and polymers [27].

4.3.2 Conductimetric immunosensor using FIDµE's

After the deposition of the biocompatible layer (SiO_2), the conductimetric immunosensor performance was analyzed using the new FIDµE's. The functionalization performed followed the covalent immobilization techniques detailed in previous sections herein above.

Firstly, an analysis of the impedimetric response was developed in order to prove that the new electrodes are able to monitoring the impedance variations related to the presence of the bioreagents. Thus, in order to qualitatively show how the immunosensor is sensitive to the atrazine concentration, experiments in assay buffer were conducted including different values of atrazine concentration (0.32 – 2000 µg L^{-1}) during the competition step using different ID's samples for every concentration. Although, as it has been previously mentioned impedimetric detection is not recommended for the conductimetric immunosensors, a curve response (Figure 12) was obtained. This curve is based in the value of the Rs in the Step V of the functionalization.

In this case, the LOD is **3.657 µg L^{-1}**, once again far below the MRL required. Nevertheless, in addition to the MRL obtained, we consider this curve as an important result because it corroborates the conclusions exposed above: in the frequency domain, the impedance contribution of the antibodies is larger than the gold particles contribution. At low atrazine concentrations, the amount of Ab$_2$ (labelled with gold particles) is larger than at high atrazine concentrations. Therefore, if the contribution of the gold particles is higher, the Rs value should be maximum at high atrazine concentrations where the amount of gold is minimum; nevertheless, the curve shows a minimum Rs value at high atrazine concentrations.

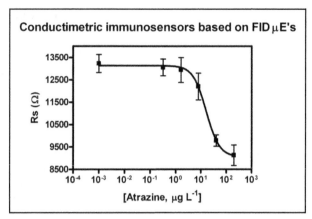

Fig. 12. Response curve of the conductimetric immunosensor, using the covalent immobilization technique, for the atrazine detection in relation with the Rs variation. Buffer solution was used for the competitive reaction. Measures were taken in diluted PBS solution. See Table 3 for the features of the atrazine assay.

Finally, the conductimetric response was also quantified. For that, the concentrations of atrazine (0.32 – 2000 µg L^{-1}) were maintained. The electrodes were covered by a diluted PBS solution and the measurements were performed with a bias of +100 mVdc. The response curve obtained can be seen in Figure 13. In this case, the curve is based on the current through the electrodes, related to amount of gold particles.

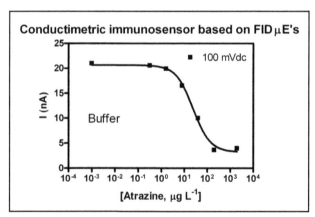

Fig. 13. Response curve of the conductimetric immunosensor, using the covalent immobilization technique, for the atrazine detection in relation with the presence of gold particles (40 nm). Buffer solution was used for the competitive reaction. Measures were taken in diluted PBS solution and the *blank* was reduced. See Table 3 for the features of the atrazine assay.

In this case, the LOD is **2.975 µg L^{-1}** and the R^2 was 0.9922. Nevertheless, an additional conclusion is that in the conductimetric case, a very small gap is not needed to achieve the atrazine detection in low concentrations.

The most relevant analytical features of the atrazine assays (impedimetric and conductimetric) are summarized in Table 3.

Features of the atrazine assays[a]	Conductimetric immunosensor (buffer)	
	impedimetric response	conductimetric response (100 mV)
IC$_{50}$, µg L^{-1}	16.14±0.16	24.17±0.03
LOD, µg L^{-1}	**3.657**	**2.975**
R^2	0.89	0.99

[a] The parameters are extracted from the four-parameter equation used to fit the standard curve.

Table 3. Features of the atrazine assays[a].

5. Pesticide residues detection in red wine samples

After the demonstration of both types of immunosensors for the detection of free pesticide in buffer samples, the immunosensors were studied using a complex matrix such as red wine samples. Red wine was chosen instead of other matrixes such as white wine, water or grape juice, because its strong matrix effect. Therefore, if the red wine matrix effect can be measured, the other matrix effects will be easier. Again, atrazine was used as pesticide of test.

5.1 Analysis of Red Wine

Red wine samples were obtained from a local retail store and used, on a first instance, to evaluate the extension of the potential non-specific interferences. Prior measurements with the immunosensors, the wine samples were purified by solid-phase extraction (SPE) using LiChrolut RP-18 (500 mg, 6 mL) sorbent (Merck, Darmstadt, Germany) pre-conditioned with MeOH (3 mL), and MeOH:Mili-Q water (15:85, v/v, 3 mL) at a flow rate of 3 mL min^{-1}. The wine samples (3 mL) were loaded at 5mL min^{-1}, and the SPE cartridges washed with of MeOH:Mili-Q water (70:30, v/v, 1 mL), dried, and finally eluted with of MeOH:Mili-Q water (80:20, v/v, 1 mL). The fractions collected were diluted 1:50 in PBST and used for the impedimetric measurements [28].

5.2 Impedimetric immunosensor

5.2.1 Detection method applied: Impedance spectroscopy

As in section 4, impedance characterization measurements in a wide frequency range and fitting of the Nyquist plots of impedance spectra to an equivalent circuit, were applied as detection method.

Using the same conditions as in the measurements of the assay buffer and in order to qualitatively show how the impedimetric immunosensor is sensitive to the atrazine concentration using real samples, experiments were carried out including atrazine concentration between 0.32 to 2000 µg L^{-1} during the competition step (Step IV) using different ID's arrays for every concentration. Thus, the response curve shown in Figure 14 was obtained using the tool of detection based in a wide range of frequency, the covalent immobilization technique, as well as real samples (red wine) as medium for the competitive assay.

The limit of detection obtained from the response curve shown in Figure 14 was **0.19 μg L⁻¹**. As in the previous case, this result is not only far below the MRL required by EC, but it also proves that atrazine can be detected with sub-ppb resolution when the competitive assay is performed in complex samples such as red wine.

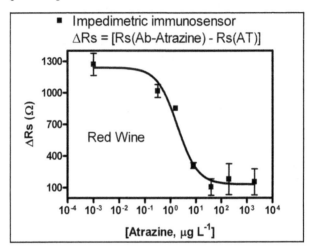

Fig. 14. Response curve of the impedimetric immunosensor, using the covalent immobilization technique, for atrazine detection in relation with the Rs variation. Real samples (red wine) were used for the competitive reaction. Measures were taken in diluted PBS solution. See Table 4 for the features of the atrazine assay.

Features of the atrazine assay	Impedimetric immunosensor (Red Wine)
Signal$_{min}$	129.7[b]
Signal$_{max}$	1239[b]
Slope	-1.095
IC$_{50}$, μg L⁻¹	1.876
LOD, μg L⁻¹	**0.19**
R²	0.86

[a] The parameters are extracted from the four-parameter equation used to fit the standard curve.
[b] Values presented are in Ω.

Table 4. Features of the atrazine assays[a].

5.3 Conductimetric immunosensor

5.3.1 Detection method applied: DC measuremnets

As in the case of the impedimetric immunosensor, the conductimetric immunosensor assay was also done using red wine samples. In order to compare results, the same conditions used previously to analyze how the conductimetric immunosensor is sensitive to the atrazine concentration using real samples were maintained. Thus, the experiments carried out in this section included the same atrazine concentrations (0.32 – 2000 µg L⁻¹) during the competition step (Step IV) using different ID's arrays for every concentration. Likewise, the electrodes were covered by a diluted PBS solution, and the measurements were performed to +25 and +100 mVdc bias.

The results obtained by this way are shown in Figure 15. Again, the *blank* was measured and reduced and the atrazine response continues to follow the inverse law which is a result of the competitive method of detection used. The limits of detection obtained for the detection of residues of atrazine, when the competitive assay was performed in red wine samples, were **0.489 µg L⁻¹** (100 mVdc bias) and **0.034 µg L⁻¹** (25 mVdc bias). Again, the MRL required by EC was largely reduced.

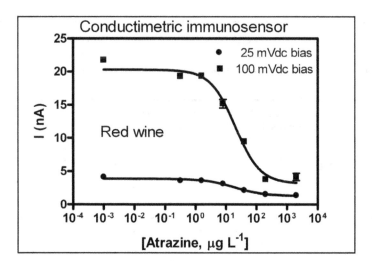

Fig. 15. Response curve of the conductimetric immunosensor, using the covalent immobilization technique, for atrazine detection in relation with the presence of gold particles (40 nm). Red wine samples were used for the competitive assay. Measures were taken in diluted PBS solution and the *blank* was reduced. See Table 5 for the features of the atrazine assay.

Features of the atrazine assays[a]	Conductimetric immunosensor (red wine)	
	25 mV	100 mV
IC_{50}, µg L^{-1}	19.05±0.10	20.54±0.07
LOD, µg L^{-1}	**0.034**	**0.489**
R^2	0.96	0.98

[a] The parameters are extracted from the four-parameter equation used to fit the standard curve.

Table 5. Features of the atrazine assays[a].

6. Immunosensors comparison

A comparison between the performance of the impedimetric and conductimetric immunosensors is shown in tables 6 and 7. From the data obtained is concluded that using either type of immunosensors, LODs far below the MRL can be obtained. These good results are directly related to the advantages of the immunosensors presented such as the use of IDµE's, the competitive assay based on the antibodies variation, as well as secondary antibody labelled with gold nanoparticles (in the case of the conductimetric immunosensor).

Features of the atrazine assays[a]	Impedimetric immunosensor	Conductimetric immunosensor
LOD, µg L^{-1}	0.04	0.104 - 1.217

[a] Limit of detection is extracted from the four-parameter equation used to fit the standard curve.
[b] Functionalization was performed by covalent immobilization techniques.

Table 6. Limit of detection of the atrazine assays performed in assay buffer [a,b].

Features of the atrazine assays[a]	Impedimetric immunosensor	Conductimetric immunosensor
LOD, µg L^{-1}	0.19	0.034-0.489

[a] Limit of detection is extracted from the four-parameter equation used to fit the standard curve.
[b] Functionalization was performed by covalent immobilization techniques.

Table 7. Limit of detection of the atrazine assays performed in red wine samples[a,b]

In theory, a labelled biosensor must be more sensitive than a label-free biosensor; nevertheless, in our immunosensors this is not always the case. For this reason, we consider that the conductimetric immunosensor performance can still be improved (in order to reach the LODs obtained for the impedimetric immunosensor). To obtain this improvement, two main facts must be taken into account:

First, it is important to remark that the label used for the conductimetric immunosensor is not enzymatic, it is a conductive label. Thus, its influence is largely related to the aspect ratio between the particle diameter (40 nm) and the electrodes gap (5000 nm). In the conductimetric immunosensor presented here, this difference is large and, because of this, the neighbourhood of each particle becomes decisive. Therefore, in order to obtain a further improvement in the LOD this difference must be reduced, for example reducing the gap between electrodes.

Second, from the chemical point of view, one of the most important differences that exist between both immunosensors is that the conductimetric one includes a second antibody. Inevitably, the inclusion of the secondary antibody affects the immunosensor performance, because the antibody has a different electrical behaviour than the gold particle linked to it. Thus, another interesting approach related the conductimetric immunosensor would be to directly include the gold nanoparticle to the first antibody, therefore eliminating the second antibody.

7. Conclusions

In this chapter, two types of immunosensors for accurate and rapid pesticide residues detection in wine samples have been explained. These devices were designed in order to meet the need of legislation that establish the limits of pesticides in wine, as well as a system which improves the control of these residues by means of the application of innovative immunodiagnostic microarray devices for rapid at line assessment of pesticides in wine processing operations.

The immunosensors presented have been named: impedimetric immunosensors and conductimetric immunosensor. The impedimetric immunosensor is based on an array of interdigitated µ-electrodes (IDµE) and bioreagents specifically developed (antigen, antibody) to detect residues of pesticide, uses unlabelled antibodies. The conductimetric immunosensor incorporates gold nanoparticles additionally. Bioreagents were covalently immobilized on the surface of the electrodes (interdigital space). In both cases the biochemical determination of pesticide is possible without any redox mediator.

For the case of the impedimetric immunosensor, the detection method is based on impedimetric measurements (in a wide range of frequencies and at single frequency), whereas in the case of the conductimetric immunosensor the detection method is based on conductimetric measurements (DC measurements).

The potential of the immunosensors to detect pesticides has been evaluated using atrazine in assay buffer as well as in red wine samples. Both immunosensors are different and each of them gives different advantages:

i.   The impedimetric immunosensor does not require the use of any label to achieve the detection of small amounts of atrazine;

ii. The impedimetric immunosensor has demonstrated the possibility of sub-ppb atrazine detection, even when the competitive assay is done in complex samples such as red wine.

iii. The conductimetric immunosensor offers the possibility of detecting small amounts (sub-ppb also) of atrazine by means of simple and inexpensive DC measurements. This opens the possibility of obtaining a commercial immunosensor of very easy use (it does not require qualified personnel), transportable and cost effective.

8. References

[1] Laureyn, W., et al., Nanoscaled interdigitated titanium electrodes for impedimetric biosensing. Sensors and Actuators B, 2000. 68: p. 360–370.

[2] Hleli, S., et al., Atrazine analysis using an impedimetric immunosensor based on mixed biotinylated self-assembled monolayer. Sensors and Actuators B, 2006. 113: p. 711-717.

[3] Helali, S., et al., A disposable immunomagnetic electrochemical sensor based on functionalised magnetic beads on gold surface for the detection of atrazine. Electrochimica Acta, 2006. 51: p. 5182–5186.

[4] Valera, E., Á. Rodríguez, and L.M. Castañer, Steady-State and Transient Conductivity of Colloidal Solutions of Gold Nanobeads. Nanotechnology, IEEE Transactions on, 2007. 6(5): p. 504-508.

[5] Cosnier, S., Biomolecule immobilization on electrode surfaces by entrapment or attachment to electrochemically polymerized films. A review. Biosensors & Bioelectronics, 1999. 14: p. 443–456.

[6] Kandimalla, V.B., V.S. Tripathi, and H. Ju, Immobilization of Biomolecules in Sol-Gels: Biological and Analytical Applications. Critical Reviews in Analytical Chemistry, 2006. 36(5): p. 73-106.

[7] Meyer-Plath, A.A., et al., Current trends in biomaterial surface functionalization – nitrogen-containing plasma assisted processes with enhanced selectivity. Vacuum, 2003. 71: p. 391–406.

[8] Siow, K.S., et al., Plasma Methods for the Generation of Chemically Reactive Surfaces for Biomolecule Immobilization and Cell Colonization - A Review. Plasma Processes and Polymers, 2006. 3: p. 392–418.

[9] Luzinov, I., et al., Epoxy-Terminated Self-Assembled Monolayers: Molecular Glues for Polymer Layers. Langmuir, 2000. 16: p. 504-516.

[10] Fredj, H.B., et al., Labeled magnetic nanoparticles assembly on polypyrrole film for biosensor applications. Talanta, 2008. 75(3): p. 740-747.

[11] Salmain, M., N. Fischer-Durand, and C.-M. Pradier, Infrared optical immunosensor: Application to the measurement of the herbicide atrazine. Analytical Biochemistry, 2008. 373: p. 61–70.

[12] Helali, S., et al., Surface plasmon resonance and impedance spectroscopy on gold electrode for biosensor application. Materials Science and Engineering C, 2007. 28(5-6): p. 588-593.

[13] Zacco, E., et al., Electrochemical biosensing of pesticide residues based on affinity biocomposite platforms. Biosensors and Bioelectronics, 2007. 22: p. 1707–1715.

[14] Anh, T.M., et al., Detection of toxic compounds in real water samples using a conductometric tyrosinase biosensor. Materials Science and Engineering C, 2006. 26: p. 453-456.

[15] Steinheimer, T.R., HPLC Determination of Atrazine and Principal Degradates in Agricultural Soils and Associated Surface and Ground Water. J. Agric. Food Chem., 1993. 41: p. 588-595.

[16] Kaune, A., et al., Soil Adsorption Coefficients of s-Triazines Estimated with a New Gradient HPLC Method. J. Agric. Food Chem., 1998. 46: p. 335-343.

[17] Moore, A. and C.P. Waring, Mechanistic Effects of a Triazine Pesticide on Reproductive Endocrine Function in Mature Male Atlantic Salmon (Salmo salar L.) Parr. Pesticide Biochemistry and Physiology, 1998. 62: p. 41–50.

[18] Papiernik, S.K. and R.F. Spalding, Atrazine, Deethylatrazine, and Deisopropylatrazine Persistence Measured in Groundwater in Situ under Low-Oxygen Conditions. J. Agric. Food Chem., 1998. 46: p. 749-754.

[19] Hou, Y., et al., Immobilization of rhodopsin on a self-assembled multilayer and its specific detection by electrochemical impedance spectroscopy. Biosensors and Bioelectronics, 2006. 21: p. 1393–1402.

[20] Hleli, S., et al., An immunosensor for haemoglobin based on impedimetric properties of a new mixed self-assembled monolayer. Materials Science and Engineering C, 2006. 26: p. 322-327.

[21] Ionescu, R.E., et al., Impedimetric immunosensor for the specific label free detection of ciprofloxacin antibiotic. Biosensors and Bioelectronics, 2007. 23: p. 549–555.

[22] Lee, J.A., et al., An electrochemical impedance biosensor with aptamer-modified pyrolyzed carbon electrode for label-free protein detection. Sensors and Actuators B, 2008. 129: p. 372–379.

[23] Kang, X., et al., Glucose biosensors based on platinum nanoparticles-deposited carbon nanotubes in sol–gel chitosan/silica hybrid. Talanta 74, 2008. 74: p. 879–886.

[24] Maggio, E.T., Enzyme-Immunoassay. 2nd edition ed. 1981, Florida: CRC Press.

[25] Valera, E., et al., Conductimetric immunosensor for atrazine detection based on antibodies labelled with gold nanoparticles. Sensors and Actuators B, 2008. 134: p. 95-103.

[26] Valera, E., et al., Determination of atrazine residues in red wine samples. A conductimetric solution. Food Chemistry, 2010. 122: p. 888-894.

[27] Yuqing, M., G. Jianguo, and C. Jianrong, Ion sensitive field effect transducer-based biosensors. Biotechnology Advances, 2003. 21: p. 527-534.

[28] Ramón-Azcón, J., et al., An impedimetric immunosensor based on interdigitated microelectrodes (IDμE) for the determination of atrazine residues in food samples. Biosensors & Bioelectronics, 2008. 23: p. 1367-1373.

4

Enantioselective Activity and Toxicity of Chiral Herbicides

Weiping Liu and Mengling Tang
Zhejiang University
China

1. Introduction

Chirality is a natural property that is well known to chemists and has been generally recognized in the life sciences since Pasteur discovered the optical isomers of tartrate and van't Hoff and LeBel proposed the theory of the stereostructure of carbon compounds (GassmannKuo & Zare, 1985; Koeller & Wong, 2001; KondruWipf & Beratan, 1998). Almost all of the biological macromolecules, such as DNA, RNA, protein, polynucleotides, and even the amino acids, the basic structural units of life, are chiral(Roelfes, 2007). Although the enantiomers of chiral substances have the same physicochemical properties, their biochemical activities, unlike abiotic transformations, can be quite different because biochemical processes usually show high stereo- or enantioselectivity (Muller & Kohler, 2004). For instance, enantioselective reactions occur in biological enrichment(Hegeman & Laane, 2002), degradation and other physiological actions(Wong, 2006). For organisms, enantiomers often exhibit different effects or toxicities. The "active" enantiomer of a chiral chemical may have the desired effect on a target species, whereas the other enantiomer may not(Garrison, 2006). It is advisable to use only the biologically active enantiomers, thereby reducing the total amount of chemical pollutants released into the environment.

Many commercial agrochemicals have chiral structures. For example, about 30% of currently registered pesticide active ingredients contain one or more chiral centers(Sekhon, 2009). Herbicides are used to control the growth of undesired vegetation, and they account for most of the agrochemicals in use today. Some chiral herbicides are sold as purified, optically active isomers, but for economic reasons, many others are still used as racemates. Different enantiomers of chiral herbicides can have different enantioselective activities on target weeds and different toxic effects on non-target organisms because of their enantioselective interactions with enzymes and biological receptors in organisms(Yoon & Jacobsen, 2003). Although the high efficiency and environmental safety of herbicides are of great concern, studies on their enantioselective activity and toxicity are still limited. It is very important to pay attention to the effects of chiral herbicides on biological systems in future research in order to achieve efficient, green, safe and pure herbicides with chirality, to make regulatory decisions and to predict the environmental risks of such herbicides.

This review summarizes the activities of different kinds of chiral herbicides that are widely used, such as phenoxyalkanoic acids, aryloxyphenoxypropionates, acetanilides, ureas, diphenyl ethers and other herbicides. It also address different types of enantioselective toxicity, including chronic toxicity, acute toxicity, and phytotoxicity. The enantioselective properties of the interactions between chiral herbicides and biological macromolecules via models in vivo or in vitro are also discussed. In further researches, finding low-cost methods for separating the enantiomers of herbicides to produce potent enantiopure herbicides, considering both the degradation and toxicity of the herbicides when assess chiral herbicides and exploring the mechanism of the interaction between chiral herbicides and receptors seem significant.

2. Enantioselective activities of chiral herbicides on target plants

The configurations of chiral herbicides are often strongly affect their biological activities. Often, only one enantiomer is target-active, or one is more target-active than the other, in which case the inactive or less active enantiomer simply contributes to the chemical load that pollutes the environment(Garrison, 2006). The activity of chiral herbicides on plants is always enantioselective because individual stereoisomers can interact differently with other chiral molecules, such as enzymes and other biological receptors(Wong, 2006), and the processes of absorption, interaction of target enzymes and metabolism are affected differently by different enantiomers. As a consequence, some enantiomers show higher activity against weeds, and others show lower activity. Because chiral herbicides have the advantages of high-efficiency and universal applicability, the technical separation of racemates or the synthesis of pure or enriched enantiomers is of growing importance for the agrochemical industry. Early in 1974, the chemical company BASF in Germany brought the enantiomerically pure chemical mecoprop-P (Fig. 1) to market, and after that, dozens of pure or enriched chiral herbicides were produced and applied successfully (Zipper, Nickel, Angst, & Kohler, 1996).

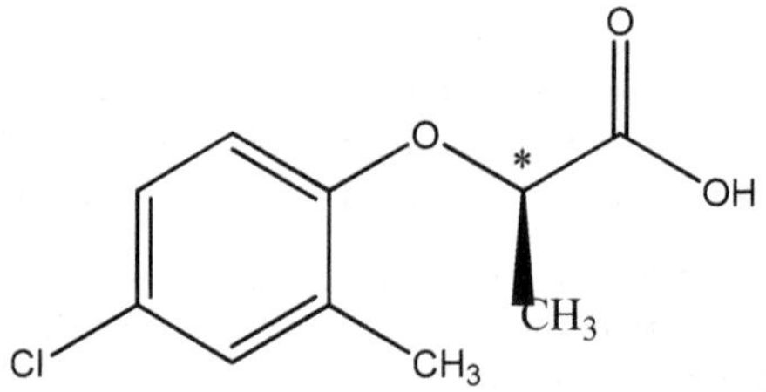

Fig. 1. Chemical structure of mecoprop-P (R-2-(4-chloro-2-methylphenoxy)propionic acid).

2.1 Phenoxyalkanoic acids

The phenoxyalkanoic acids (Fig. 2) comprise an important set of organic compounds, of which several halogenated analogues are commercially available as auxin or "hormone" herbicides(Kennardsmith & White, 1982). They are widely used to control broadleaf weeds in agriculture, lawn care, and industrial applications. Phenoxyalkanoic acid herbicides work by inhibiting acetyl-CoA carboxylase(ACCase), which leads to the termination of fatty acid synthesis, abnormal cell growth and division, and, ultimately, suppression of weed growth.

The most significant chiral compounds that have been commercialized are mecoprop and dichlorprop. Mecoprop and dichlorprop are chiral molecules that each have one stereogenic center and, therefore, exist as two enantiomers(Muller, Fleischmann, van der Meer, &

Kohler, 2006). As early as 1953, it was reported that the R-enantiomers have herbicidal activity, and the S-enantiomers have little observable activity(Matell, 1953). Authorities in the Netherlands and Switzerland have revoked registrations for racemic mixtures of chiral phenoxy herbicides while approving registrations of single-isomer products (named mecoprop-P and dichlorprop-P)(W. P. Liu, J. Ye & M. Q. Jin, 2009). In plants, mecoprop and dichlorprop were found to be enantioselectively degraded in a study conducted by Schneiderheinze. The S-(-)-enantiomer of each herbicide was preferentially degraded in most species of broadleaf weeds, whereas the R-(-)-enantiomer of each herbicide was more resistant to degradation(Schneiderheinze, Armstrong & Berthod, 1999).

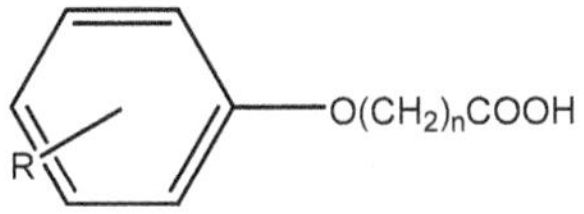

R=phenoxy, aryloxy or aromatic heterocyclicoxyl; n=1, 2 or 3.
Phenoxyalkanoic acids

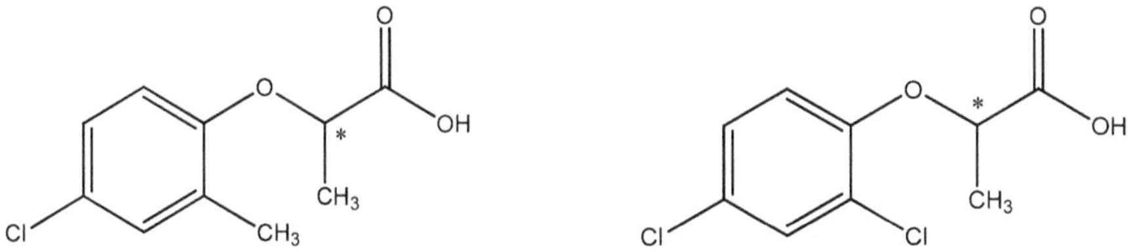

Mecoprop
(2-(R,S)-2-methyl-4-chlo-rophenoxypropanoic acid

Dichlorprop
(2-(R,S)-2,4-dichlo-rophenoxypropanoic acid)

Fig. 2. Chemical structures of phenoxyalkanoic acids*; asymmetric carbon.

2.2 Acetanilides

The chirality of acetanilide herbicides (Fig. 3) is caused by the presence of two asymmetric carbons and two chiral axes that generate two pairs of enantiomers, and the herbicide activity is mainly attributed to the S-enantiomer of the carbon in the alkyl substituent. The S-enantiomers of the herbicides metolachlor (Schmalfuss, Matthes, Knuth, & Boger, 2000) and dimethenamide (Couderchet, Bocion, Chollet, Seckinger, & Boger, 1997; Gotz & Boger, 2004) were reported to inhibit fatty acid synthase. Using acyl-CoA as a substrate to test the activity of the chiral chloroacetamide metolachlor, it was shown that only the herbicidally active S-enantiomer could inhibit elongation, whereas the R-enantiomer had no effect (Schmalfuss et al., 2000). When 5 μmol/L dimethenamid S-enantiomer was applied, the algal growth and fatty acid desaturation were strongly inhibited, but the R-enantiomer had almost no effect on algal growth (Couderchet et al., 1997). Also, the inhibition of protein synthesis and RNA polymerase I activity was found to occur as part of the active mechanism of acetanilides (Chesters et al., 1989; Liu et al., 2009).

Metolachlor is a widely used herbicide that inhibits the synthesis of fatty acids in broadleaf weeds. In 1982, it was found that the two 1S stereoisomers of metolachlor provide most of its biological activity. The herbicidal activity of the S-enantiomers was almost 10 times higher than that of the R-enantiomers (Blaser et al., 1999; Fayez & Kristen, 1996). Their activity is mainly influenced by the configuration at the chiral centre, a carbon in one of the alkylic substituents of the nitrogen in the imide group, and by the atropisomerism generated by the hindered rotation around the aryl carbon-nitrogen bond (Polcaro et al., 2004). A systematic experiment found that acyl-CoA elongation was only inhibited by the herbicidally active S-enantiomer, whereas the R-enantiomer had no influence. Furthermore, enzyme activity could not be recovered by dilution of the enzyme-inhibiting chiral herbicide(Schmalfuss et al., 2000).

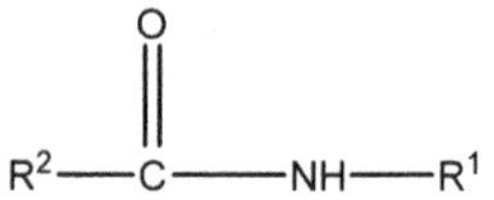

Acetanilides

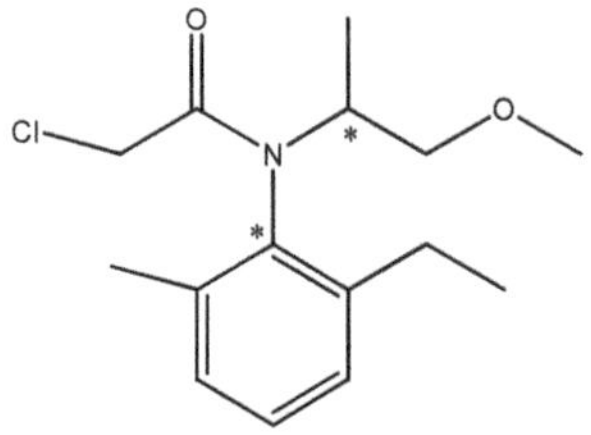

Metolachlor	Dimethenamide
(2-chloro-N-(2-ethyl-6-methylphe-nyl)-	[2-chloro-N-(2,4-dimethyl-3-thienyl)-
N-(2-methoxy-1-methylethyl)acetamide)	N-(2-methoxy-1-methylethyl)acet-amide]

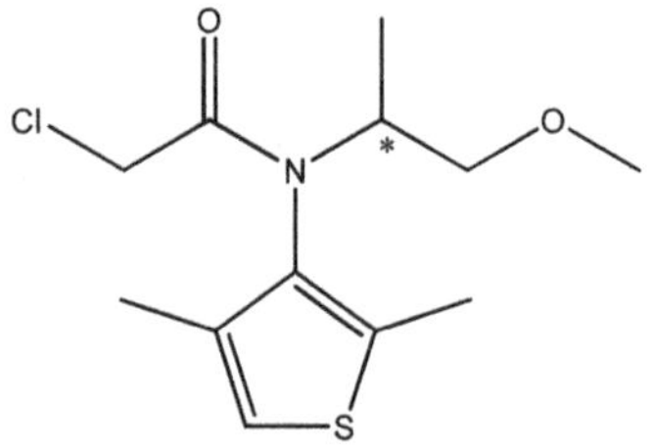

Fig. 3. Chemical structures of acetanilides.

2.3 Aryloxyphenoxypropionates

Aryloxyphenoxypropionates (AOPP) are postemergence herbicides that cause almost immediate growth inhibition in the shoot, root and intercalary meristems. Diclofop-methyl, fluazifop-P, haloxyfop-methyl, fenoxaprop-P-ethyl, quizalofop-ethyl, Fenthiaprop, and fenoxaprop-P- ethyl (Fig. 4) are all examples of chiral AOPP herbicides. A wide variety of AOPPs and their esters have been developed as commercial herbicides. Recent studies indicate that the mechanism controlling the growth of grasses by this kind of herbicide is the same as that of phenoxyalkanoic acids: they interfere with lipid metabolism in susceptible plants and inhibit the plastid ACCase, a key enzyme in long-chain fatty acid biosynthesis (Kunimitu et al., 1988; Liu et al., 2009). Chiral AOPPs have enantioselective activity on target plants. Their herbicidal activity comes almost entirely from the R-enantiomers rather than the S-enantiomers, which means that the R-enantiomers are more effective than the S-enantiomers for weeding (Sakata et al., 1985). Many reports have described the enantioselective activity of AOPPs on target weeds.

Both enantiomers of diclofop-methyl show similar pre-emergence herbicidal activity for controlling weeds in the rice field, but in postemergence applications, the R-(+)-isomer has higher activity against millets and oats. The most likely mechanisms of action for diclofop-methyl were discussed in the context of its role in oxidative membrane catabolism by free radical lipid peroxidation and its coupling to the effect of diclofop on the transmembrane proton gradient (Kurihara et al., 1997; Shimabukuro & Hoffer, 1995). Racemic mixtures and

(+)-AOPP are active in alfalfa embryo induction, whereas the (-)-forms are inactive and do not inhibit embryogenesis (Stuart & Mccall, 1992).

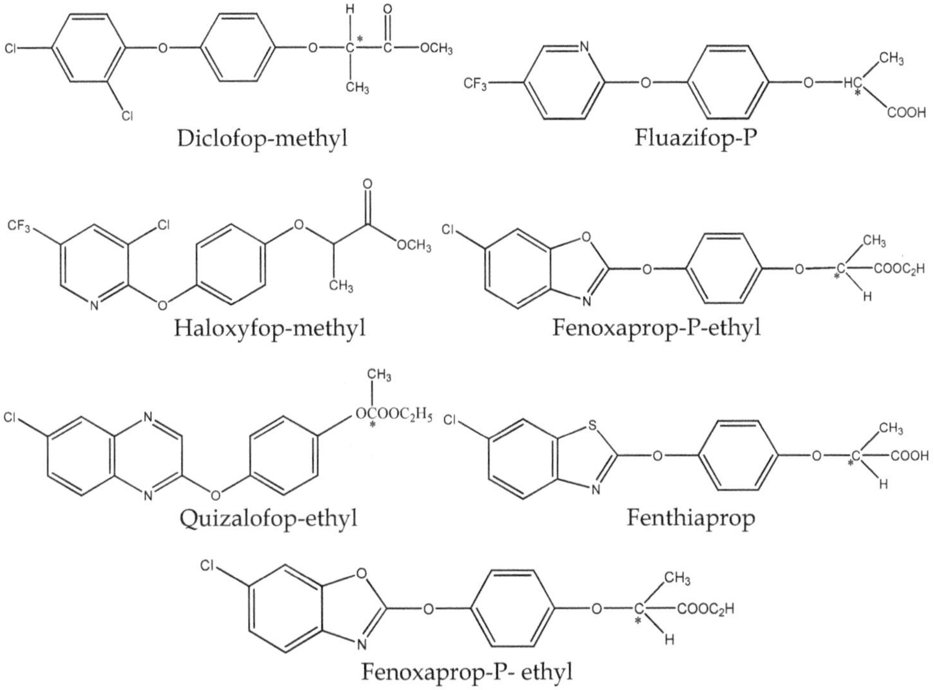

Diclofop-methyl   Fluazifop-P

Haloxyfop-methyl   Fenoxaprop-P-ethyl

Quizalofop-ethyl   Fenthiaprop

Fenoxaprop-P- ethyl

Fig. 4. Chemical structures of aryloxyphenoxypropionates.

2.4 Ureas

The key structural feature of substituted urea herbicides (Fig. 5) is the urea moiety, and different substituents of the amino groups produce various kinds of urea herbicides. The action of this kind of herbicide depends on differences in absorption, conduction and degradation abilities between plants and weeds. Ureas act by inhibiting the Hill reaction in photosynthetic electron transport (Jurado et al., 2011). Cycluron, daimuron and clodinafop-propargyl are the main brands of urea herbicides. Of these, daimuron and R-clodinafop-propargyl are the ones that have been commercialized. Daimuron has a very plant-specific effect: it shows herbicidal activity against paddy weeds. The R-enantiomer inhibits the growth of *Cyperaceae* weeds more strongly than the S-enantiomer (Ryoo et al., 1998; Omokawa et al., 1999). The herbicide 1-α-methylbenzyl-3-p-tolyl urea (MBTU), a derivative of daimuron, was shown to have enantioselective differences in potency. For instance, it was reported that the enantiomers of MBTU have different depression effects on roots in a number of *Oryzeae*, *Echinochloa* and wheat species, and the root growth of all members of the genus *Oryza* was inhibited more strongly by R-MBTU than by S-MBTU. In contrast, the root growth of *Echinochloa* and wheat was more sensitive to S-MBTU than to the antipodal R-MBTU (Omokawa et al., 2004; Kazuhiro et al., 2009). Imai et al. thought that the decrease in free amino levels in the root tips was the reason that R-/S-MBTU inhibited the growth of the plants (Imai, Kojima & Numata, 2009). The activity of α-methylbenzyl-p-tolylureas (4-Me) mainly depends on the substituents of benzene. The 4-Me *R*-enantiomers with a smaller alkyl group exhibited significant activity on both of the plant species on which they were tested (Omokawa & Ryoo, 2001).

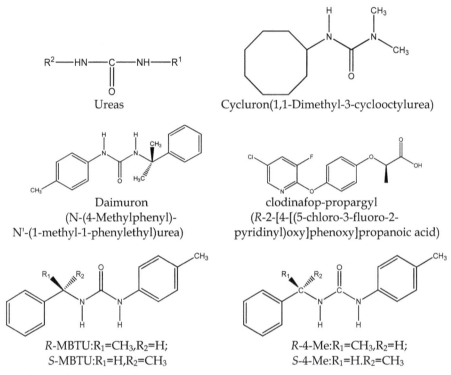

Fig. 5. Chemical structures of ureas.

2.5 Diphenyl ethers

Diphenyl ethers (Fig. 6) can inhibit the photosynthesis and affect the composition of chloroplasts, thereby causing the death of weeds. Among diphenyl ether herbicides, nitrodiphenyl ethers have a chiral structure and are used to control broadleaf weeds. They are able to cause light-dependent membrane lipid peroxidation, and their *S*-(-)-enantiomers have been shown to be substantially more active than their *R*-(+)-enantiomers in a test designed to monitor plant membrane breakdown by Camilleri et al. This finding presumably reflected the fact that the binding of nitrodiphenyl ethers to a metabolic enzyme in plants is enantioselective. Nitrodiphenyl ethers and their analogues act by increasing the level of protoporphyrin IX in an enantiotopically specific active site by inhibiting an enzyme in the biosynthetic pathway between protoporphyrin IX and protochlorophyllide (Camilleri et al., 1989). A test using the green alga *Chlamydomonas reinhardtii* as an indicator species showed that 5-[2-chloro-4-(trifluoromethyl)phenoxy]- 3-nitroacetophenone oxime-*O*-(acetic acid, methyl easter) (DPEI), also a type of nitrodiphenyl ether herbicide, had enantioselective activity. The purified *S*-(-)-enantiomer had greater herbicidal activity than the R-(+)-isomer, and the mechanistic reason was that the S-(-)-enantiomers of DPEs had greater potency in inhibiting protoporphyrinogen IX oxidase (Hallahan, Camilleri, Smith, & Bowyer, 1992).

Lactofen, a diphenyl ether, has an asymmetrically substituted C atom and comprises a pair of enantiomers; the herbicidal activity mostly comes from the S-(+)-enantiomer. This herbicide is applied as a foliar spray on target weeds and is used to control broadleaf weeds in soybeans, cereal crops, potatoes, and peanuts (Diao et al., 2009).

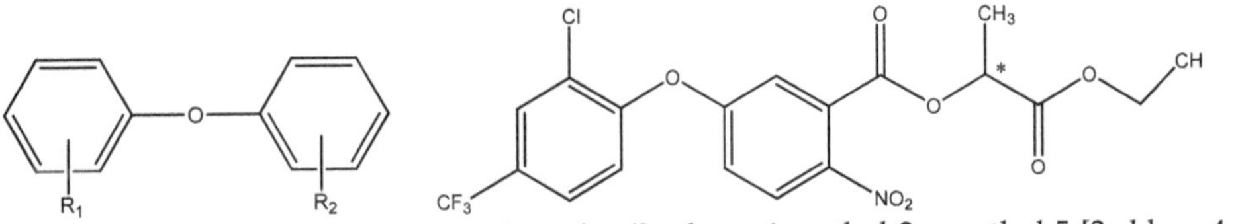

Diphenyl ethers

Lactofen (2-ethoxy-1-methyl-2-oxoethyl-5-[2-chloro-4-(trifluoromethyl)phenoxy]-2-nitrobenzoate)

Fig. 6. Chemical structures of diphenyl ethers.

2.6 Other chiral herbicides

Some organophosphorus compounds (Fig. 7) are used as herbicides. In *in vitro* activity studies, the active site of acetylcholine esterase (AChE) may interact differently with the different enantiomers of these compounds, but enantioselective differences in metabolic detoxification or toxicity may also be a major factor in determining the activity of organophosphorus herbicides.

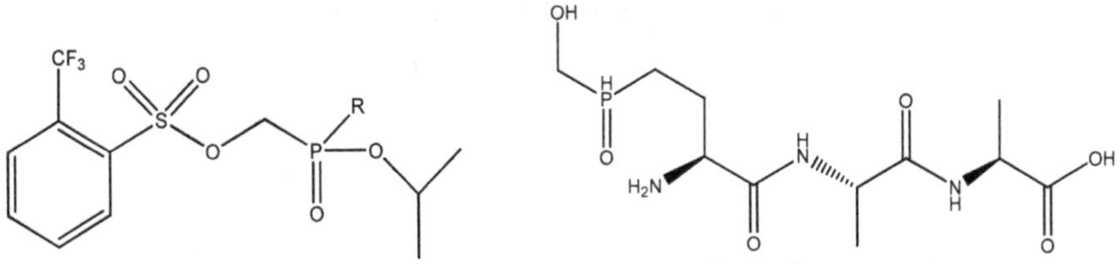

Phosphosulfonates

Bialaphos

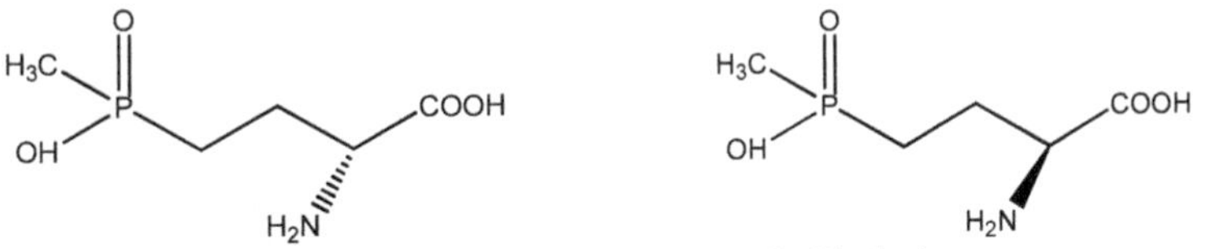

D-Glufosinate

L-Glufosinate

Fig. 7. Chemical structures of organophosphorus herbicides.

Phosphosulfonates are a class of soil-applied herbicides with activity against a variety of grassy weeds, and their chirality is attributed to an asymmetrically substituted phosphorus atom. Biological testing of the enantiomeric phosphosulfonate herbicides demonstrated that the purified (+)-enantiomer is more active than the racemate (Spangler et al., 1999). Bialaphos, another organophosphorus herbicide, has carbon chiral centres, and its S-(+)-

isomer is more active as a herbicide than its R-(-)-isomer. Glufosinate, also an organophosphorus herbicide, was used as its ammonium salt, and the activity of its enantiomers was studied using cell culture in several plant species. The results illustrated that the glufosinate racemate and L-glufosinte are transformed into the same metabolites, but D-glufosinate is not metabolized (Muller, Zumdick, Schuphan, & Schmidt, 2001; RuhlandEngelhardt & Pawlizki, 2002).

Certain triazines (Fig. 8) are often used as plant growth regulators. The ones having a chiral nitrogen center, such as amitrole, atrazine, cyanazine, and simazine, are effectively achiral because their enantiomers can easily be interconverted. The triazines with chiral C centres are used as selective weed killers and act by inhibiting photosynthesis (W. LiuJ. Ye & M. Jin, 2009).

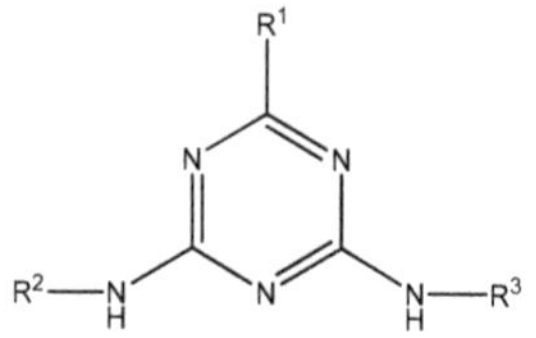

Fig. 8. Chemical structures of triazines.

All imidazolinone herbicides (Fig. 9) are chiral, and imazapyr, imazapic, imazethapyr, imazamox and imazaquin are widely used herbicides from the imidazolinone family (Lao & Gan, 2006). It has been reported that their R-enantiomers are 10 times more inhibitory toward the enzyme acetolactate synthase (ALS) than their S-enantiomers (Chin, Wong, Pont, & Karu, 2002). Imazethapyr (IM) is always absorbed through the roots of plants, and Zhou et al. found that R-(-)-IM affected the root growth of maize seedlings more severely than S-(+)-IM (Zhou, Xu, Zhang, & Liu, 2009).

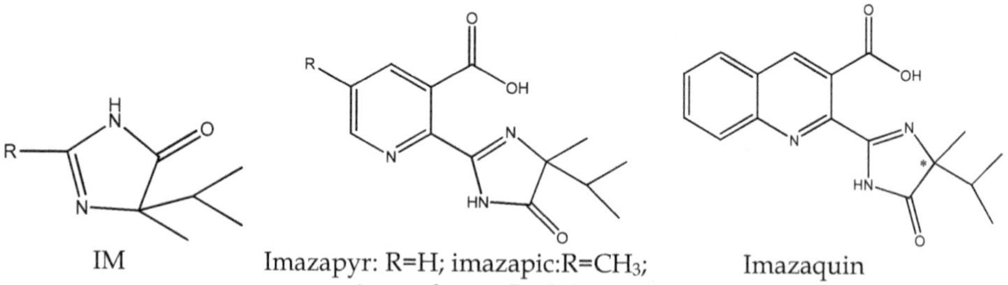

IM Imazapyr: R=H; imazapic:R=CH$_3$; imazethapyr:R=CH$_2$CH$_3$; imazamox: R=CH$_2$OCH$_3$ Imazaquin

Fig. 9. Chemical structures of imidazolinone herbicides.

Cyclonenes (Fig. 10), which are used to control broadleaf weeds and can be absorbed by the leaves of plants, can inhibit the synthesis of fatty acids by acting on acetyl-CoA carboxylase. Among cyclonenes, clethodim is a selective post-emergence herbicide for the control of annual and perennial grasses. The optically pure (-)-enantiomer of clethodim was, surprisingly, more effective in regulating the growth of grass plants than the corresponding racemic mixture or the optically pure (+)-enantiomer (Whittington et al., 2001).

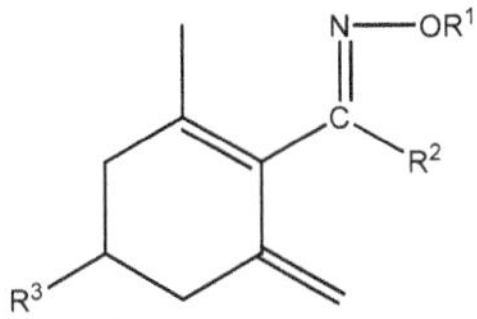

Cyclonenes 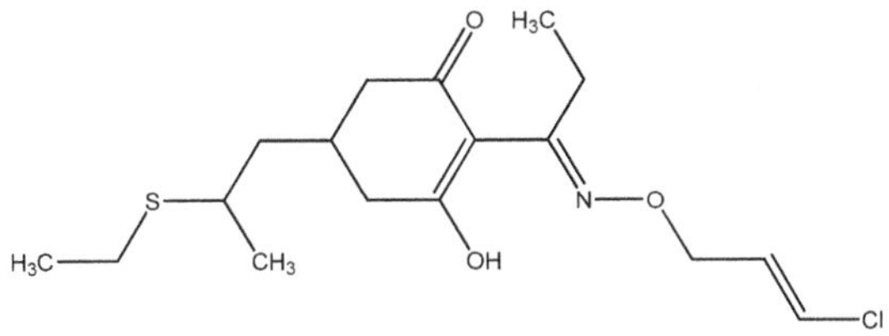   clethodim

Fig. 10. Chemical structures of cyclonenes.

3. Enantioselective toxicity of chiral herbicides on non-target organisms

Different enantiomers of chiral herbicides can have different potencies on their target plants, and they also may have enantioselective toxic effects on non-target organisms. In previous studies, researchers always focused on the efficacy of chiral herbicides while neglecting the negative biological effects associated with particular enantiomers that might persist in the environment long after application. The potential biological toxicities of these herbicides, which could include chronic toxicity, acute toxicity and phytotoxicity, are generally enantioselective. When the Environmental Protection Agency (EPA) developed its assessment method for acute toxicity testing using *Daphnia* and *Ceriodaphnia dubia* as aquatic indicators, many researches around the world undertook studies of the enantioselective toxicity and molecular mechanisms of chiral herbicides using both *in vivo* and *in vitro* models. Lewis et al. found that all pasture samples from Brazilian soils preferentially transformed the non-herbicidal enantiomer of dichlorprop via a microbial transformation processes, whereas most forest sample transformed the herbicidal enantiomer more rapidly or as rapidly as the non-herbicidal enantiomer (Lewis et al., 1999). In general, the faster the microbial degradation processes remove the herbicides from the environment, the safer the effect of the herbicides is. Hence, using a pure enantiomer that is more active and safe than its partner or racemate is likely to cause less environmental damage and to pose less ecological risk (Hegeman & Laane, 2002).

3.1 Enantioselectivity in the chronic toxicity of chiral herbicides

For chiral pesticides, there have been several reports describing enantioselective chronic toxicity, but there are few such reports about herbicides. They are limited to a small number of reports about metolachlor. Zhan and Xu used silkworm as a biological indicator to investigate the enantioselective toxicity of metolachlor as estimated by the metabolism and activities of silkworm. Their researches mainly focused on enzyme activities, and the results suggested that rac-metolachlor and *S*-metolachlor have different effects on enzyme activities of fifth-instar silkworms (*Bombyx mori* L.). Acid phosphatase (ACP) activity in silkworm hemolymph was 44-73% higher in control organisms than in those treated with rac-metolachlor, but there was no great difference between *S*-metolachlor treatment and the control. Hemolymph lactate dehydrogenase and catalase activities were much lower in rac-metolachlor-treated silkworms than in *S*-metolachlor-treated ones. For midgut alkaline phosphatase activity, the activity found in controls was greater than in those treated with rac-enantiomers, which in turn was greater than the activity found in *S*-metolachlor-treated silkworms(Zhan, 2006). Changes in avoidance behaviour, body weight and *in vivo* enzyme activity were found in earthworms (*Eisenia foetida, E.foetida*) studied by Xu et al. When the same treatment concentrations were used, the effects of rac-metolachlor on the enzyme activities and body weight of *E.foetida* were more significant than those of *S*-metolachlor. In 2 days and 7 days' experiments, the effects of the *S*-enantiomer on cellulase and catalase activities occurred more quickly, but over the long term (14 days, 28 days), the rac-enantiomer had greater toxic effects. The test of avoidance behaviour shows that earthworms are more sensitive to the stimulation of rac-metolachlor than that of *S*-metolachlor(Xu et al., 2010) . Those two studies indicate that the rac-metolachlor is more toxic to economically important silkworms than *S*-metolachlor, and they show that the metabolic inhibition is mediated by an inhibitory effect on enzyme activity.

In a chronic toxicity test of rac-metolachlor and *S*-metolachlor, the lowest-observed-effect concentration (LOEC), no-observed-effect concentration(NOEC), number of days to first brood, length, longevity, number of broods per female, number of young per female, and the intrinsic rate of natural increase were determined using *Daphnia magna* as an indicator. The LOEC and NOEC of the rac-enantiomers were much lower than those of the *S*-enantiomers, and the longevity and number of broods per female were significantly affected when the rac-enantiomer concentration was higher than 1.0 mg L^{-1} or when the *S*-enantiomer concentration was higher than 10 mg L^{-1}. Also, the number of broods per female and the intrinsic rate of natural increase were significantly reduced when the rac-enantiomers concentration was higher than 0.01 mg L^{-1} or the *S*-enantiomers concentration was higher than 0.5 mg L^{-1}. Body length was affected by both of the herbicides, but the number of days to first brood was not affected (Liu, Ye, Zhan, & Liu, 2006). These results were in agreement with the earthworm experiments described above and showed that the chronic toxicity of metolachlor is significantly higher than that of *S*-metolachlor.

3.2 Enantioselectivity in the acute toxicity of chiral herbicides

Early studies on the enantioselective toxicity of chiral pesticides primarily tested their acute toxicity in living organisms. Dramatic differences between enantiomers were observed in tests of acute toxicity toward terrestrial and freshwater invertebrates, suggesting that the acute toxicity is primarily attributable to a specific enantiomer in the racemates. In a study using *Chlorella pyrenoidosa* to test the acute toxicity of rac-metolachlor and *S*-metolachlor, the growth inhibition rate, chlorophyll *a* and chlorophyll *b* concentrations, catalase activity and ultrastructural morphology of cells were used as toxicity endpoints. The values of the 24, 48, 72, and 96 h EC_{50} and the chlorophyll *a* and chlorophyll *b* concentrations measured in the organisms exposed to rac-metolachlor were higher than those obtained when the *S*-enantiomer was used. The catalase activity of *C. pyrenoidosa* after treatment with *S*-metolachlor for 96 h was higher than after treatment with rac-metolachlor (Liu & Xiong, 2009). The acute 24-h LC_{50} of rac- and *S*-metolachlor were also assayed in *D. magna*, and the results showed that the rac-metolachlor LC_{50} was higher than that of the *S*-isomer (Liu et al., 2006). Both of these studies indicated that *S*-metolachlor was more toxic to aquatic organisms than rac-metolachlor. Another study of acute toxicity employed a standard OECD filter paper test, an artificial soil test and a natural soil test, and this study found that there were almost no enantioselective differences in the LC_{50} for earthworms, indicating that the acute toxicities of the two chiral herbicides displayed no enantioselectivity(Xu et al., 2009).

The dissipation and degradation of the herbicide lactofen are enantioselective processes: under laboratory conditions using enantioselective HPLC, the S-(+)-enantiomer is degraded faster than the R-(-)-enantiomer, producing residues enriched with R-(-)-lactofen when the racemic compound is incubated under aerobic and anaerobic conditions in sediments (Diao et al., 2009). The enantioselective acute toxicity of individual enantiomers of lactofen and its metabolite, desethyl lactofen, were studied in D. *magna*. The observed LC_{50} values of S-(+), rac-, and R-(-)-lactofen were 17.689, 4.308, and 0.378 µg/mL, respectively, and the corresponding values for desethyl lactofen showed a similar pattern (Diao et al., 2010). Therefore, the preferential degradation of S-(+)-lactofen leads to a higher concentration of the R-(-)-enantiomer, which has been shown to have higher acute toxicity to the non-target organism D. *magna*.

The acute toxicity of a series of organophosphorous compounds (OPs), 1-(substituted phenoxyacetoxy)alkylphosphonates (Fig. 11), which contain a chiral carbon atom, was also studied, and these compounds also display enantioselectivity. In an aquatic toxicity test using D. *magna* as an indicator, the *in vivo* assays showed that there is a significant difference in LC_{50} between the two enantiomers of O,O-dimethyl-1-(4-chlor- ophenoxyacetoxy) ethylphosphonate, with the (+)-enantiomer being 8.08 times more toxic than the (-)-form. Although the difference between the enantiomers of the other compounds in this study is not remarkable (1.2 to 4.2-fold), it can nevertheless be inferred that the toxicities of most chiral OPs are enantioselective (Li et al., 2008).

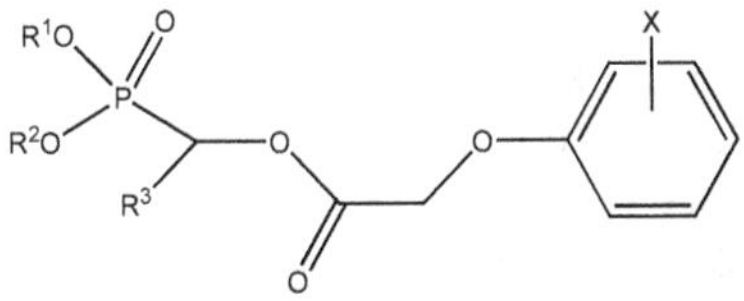

Fig. 11. Chemical structure of 1-(substituted phenoxyacetoxy)alkylphosphonates.

3.3 Enantioselectivity in the phytotoxicity of chiral herbicides

Many herbicides have been shown to be toxic to non-target plants, but our understanding of the enantioselective phytotoxicity of chiral herbicides is still limited. The chiral herbicide dichlorprop-methyl (2,4-DCPPM, Fig. 12) was clearly shown to have enantioselective toxicity toward *Chlorella pyrenoidosa*, *Chlorella vulgaris* and *Scenedesmus obliquus*. The rank order of enantiomer phytotoxicity was given by R-2,4-DCPPM > S-2,4-DCPPM > rac-2,4-DCPPM, and the toxicity of R-2,4-DCPPM was found to be about 8-fold higher than that of rac-2,4-DCPPM. All three algae species degraded 2,4-DCPPM quickly, but extraordinarily, rac-2,4-DCPPM was preferentially degraded by *Scenedesmus obliquus* at a much faster rate than the S- or R-enantiomers alone (racemate>R->S-), such that the racemate showed low toxicity compared to the other enantiomers. This phenomenon might occur because the R- and S-enantiomers are not hydrolyzed in the first 12 hours, and hydrolysis proceeds slowly after 12 hours (Li, Yuan, Shen, Wen, & Liu, 2008).

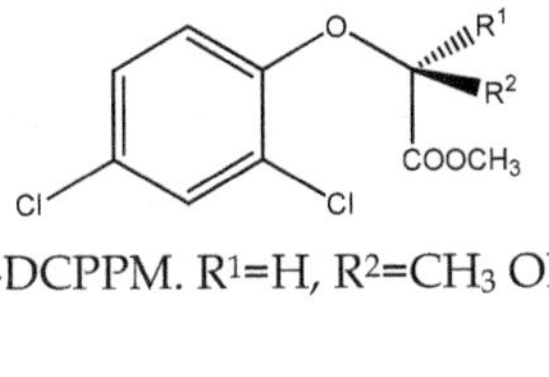

Fig. 12. Chemical structure of 2,4-DCPPM. R^1=H, R^2=CH$_3$ OR R^1=CH$_3$, R^2=H.

Several studies focused on the enantioselective phytotoxicity of IM, an herbicide widely used because of its low application rate, low toxicity to animals and broad spectrum of weed control activity (TanEvans & Singh, 2006). Zhou et al. evaluated the phytotoxicity of IM on the roots of maize (*Zea mays* L.) seedlings. Plant growth measurements and morphological, microscopic, and ultrastructural observations were conducted after treatment with individual IM enantiomers and the racemate. Although the different enantiomers showed the same trend of effects on indicators, R-(-)-IM affected the root growth of maize seeding more severely than S-(+)-IM (Zhou et al., 2009). Another study used seedlings of Xiushui 63, a Japonica rice variety to evaluate the phytotoxicity of IM; in this study, rice seedling morphology, antioxidant enzyme activity, oxidation markers and gene transcription were used as endpoints. Different enantiomers of IM also showed the same trend of effects on the seedling morphology of rice, but the levels of inhibition observed in roots and shoots showed enantioselectivity. The maximal root relative inhibition and shoot relative inhibition were ranked as follows: R-(-)-IM > racemate > S-(+)-IM. The activities of SOD, POD, and CAT and the MDA content in plants treated with R-(-)-IM were higher than those in plants treated with S-(+)-IM. In seed tissue and shoot tissue, it was observed that R-(-)-IM inhibited gene transcription and mRNA expression more strongly than S-(+)-IM (Qian et al., 2009). Both of these studies concluded that R-(-)-IM was more toxic than S-(+)-IM.

Diclofop acid (Fig. 13), produced by hydrolysis of diclofop methyl, is an herbicidal form of diclofop methyl. Significant differences were observed between its two enantiomers in an acute toxicity (72 h EC_{50}) test using rice *Xiushui* 63 seedlings. The S-enantiomer showed stronger toxicity to leaves, and the R-enantiomer was found to be more toxic to roots. The Hill reaction activity test indicated that the two enantiomers had enantioselective effects on chloroplasts, but the effects were quite complex and needed further interpretation (Ye, Zhang, Zhang, Wen, & Liu, 2009). The herbicidally inactive S-(-)-enantiomers of both diclofop-methyl and diclofop have similar or higher toxicity than the R-(+) forms to algae, depending on the algal species used. Cai et al. showed that both rac-diclofop and R-diclofop decrease algal cell permeability and that the R-enantiomer shows stronger inhibition. In contrast, the S-enantiomer increases algal cell permeability when low treatment concentrations are used, and it reduces algal cell permeability to at lesser extent only at higher concentrations compared to the R-enantiomer and rac-diclofop. The enantioselective degradation of diclofop in algae cultures is controlled by the facilitated uptake by algae, whereas the enantioselective toxicity is primarily governed by passive uptake (CaiLiu & Sheng, 2008).

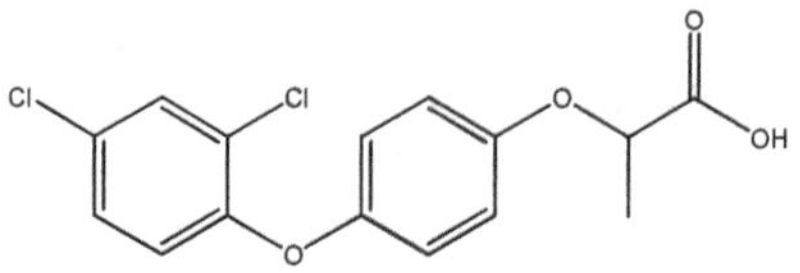

Fig. 13. Chemical structure of diclofop acid (R,S-2-[4-(2,4-dichlorophenoxy) phenoxy] propanoic acid)

4. Causative mechanisms of enantioselective herbicidal effects

The enantiomers of a chiral herbicide possess different biological activities, and one of the enantiomers usually shows a higher level of activity or toxicity. The enantiomers can alter the activities and conformations of enzymes, thereby influencing their functions and metabolic effects. Therefore, the exploration of the causative mechanisms of enantioselective effects is regarded as one of primary goals of biological chemistry. Different receptors are enantioselectively affected by herbicides for different mechanistic reasons. Wen et al. found that the enantioselective behaviours of chiral compounds might change during interactions with different chiral receptors that coexist in different biological environments. They used UV differential spectrophotometry and fluorescence spectrophotometry to study the enantioselective interactions between *Penicillium expansum* alkaline lipase and dichlorprop (DCPP) herbicide. The ranking of dichlorprop compound interaction strengths with lipase was as follows: *R*-enantiomer > rac-DCPP >*S*-enantiomer. The lipase-catalyzed kinetic experiments proved that a hydrophobic interaction seemed to play a dominant role in the interactions, and they showed that the *R*-enantiomer inhibits lipase more severely, possibly due to its stronger interaction with lipase (Wen, Yuan, Shen, Liu, & Liu, 2009). To further study this interaction, the authors conducted several tests to evaluate the toxicities toward green algae of DCPP compounds and their complexes with chitosan molecules (DCPP-CS) and chitosan nanoparticles (DCPP-NP). The results showed that, without other chiral molecules, *S*-DCPP was more toxic to *Chlorella vulgaris* than *R*-DCPP, whereas to *Scenedesmus obliquus* and *Chlorella pyrenoidosa*, *R*-DCPP was more toxic. While in the presence of CS or NP, the chiral selectivity of DCPP could be changed. For instance, the order of inhibition to *Chlorella vulgaris* was as follows: *R*-DCPP-CS > *S*-DCPP-CS and *R*-DCPP-NP > *S*-DCPP-NP. This order was the complete opposite of that observed for *Scenedesmus obliquus* and *Chlorella pyrenoidosa* (Wen, Chen, Yuan, Xu, & Kang, 2011; Wen et al., 2010). Fluorescence spectroscopic analysis showed that the interaction between CS and DCPP enantiomers depends greatly on the steric structure of DCPP. A highly stereospecific interaction between herbicide and enzyme is thought to be the typical mechanism of enantioselectivity for chiral herbicides. The three-point model proposed by Easson and Stedman indicates that, when three ligands of an herbicide match three chiral locations in the active part of an enzyme, the herbicide will show maximum potency (Easson & Stedman, 1933). If the binding sites on the protein are in a cleft or on protruding residues, the four-location model developed by Mesecar should be considered (Zhou, Liu, Zhang, & Liu, 2007). For instance, IM is an ALS-inhibiting chiral herbicide. Qian et al. investigated the enantioselectivity of *R*- and *S*-IM in *Arabidopsis thaliana*. The result showed *R*-IM powerfully induced reactive oxygen species (ROS) formation while drastically reducing antioxidant gene transcription and enzyme activity, resulting in oxidative stress. This led to the accumulation of glucose, maltose and sucrose in the cytoplasm and chloroplast, and it disrupted the carbohydrate metabolism. This result proved that enantioselectivity also affects starch metabolism in *Arabidopsis thaliana* (Qian et al., 2011).

5. Further research opportunities

The activity and toxicity of chiral herbicides should be investigated at the chiral level because enantiomers of herbicides are known to selectively interact with biological molecules that are usually enantioselective and may behave as drastically different compounds. This enantioselectivitiy varies depending on the species of biological receptor; one enantiomer of an herbicide may inhibit the growth of a particular plant while stimulating the growth of other plants. A particular enantiomer of an herbicide may be more effective than the racemate, such that using enantiopure herbicides could increase their potency for weeding and reduce the environmental burden applied chemicals. Techniques for separating the enantiomers of certain herbicides, such as chiral HPLC, GC columns (PirkleLee & Welch, 1997) and chiral electrophoresis(Desiderio, et al., 1997) , have been reported, but for economic reasons, many chiral herbicides with enantioselective activity are still used as racemic mixtures. It is important to find low-cost methods for separating the enantiomers of herbicides, which may require finding more effective chemical or biological catalysts for synthesising enantiopure herbicides.

Many previous studies have shown that, for some chiral herbicides, the active enantiomer may be environmentally degraded faster than the inactive enantiomer or the racemate, and the inactive enantiomer may even have stronger toxicity to non-target organisms. Thus, the abundance and high toxicity of the inactive enantiomer may produce passive effects in the environment. Considering the herbicidal enantioselectivity of degradation or toxicity alone would have limited environmental significance. It is important to consider both the degradation and toxicity of an herbicide enantiomer when predicting the environmental effects of the herbicide.

Herbicide activity at certain sites in non-target organisms has been reported. For instance, quizalofop and haloxyfop may inhibit ACCase, disrupt lipid metabolism and interfere with membrane transport. The potential enantioselective effects of chiral herbicides in these processes remain poorly understood and should be further explored. Though the activities and toxicities of many kinds of isolated enantiomers have been tested and reported, our understanding of the conversions between different enantiomers *in vivo* are still limited. One enantiomer may exhibit certain effects on biological receptors, but it may have opposite effects on the same target receptor when it changes into another enantiomer or racemate *in vivo*. Elucidating the transformation mechanisms and processes is a significant goal for further research. It seems that there is still a severe lack of knowledge about the characteristics and metabolism of chiral herbicides.

6. Acknowledgment

We are deeply indebted to a highly talented group of co-workers and other scientists whose names are in the relevant references. We are also thankful for financial support from the National Basic Research Program of China (2010CB126101), and the National Natural Science Foundations of China (Nos. 21177112, 40973077, 20837002, 30771255 and 20225721).

7. References

Blaser, H. U., Buser, H. P., Coers, K., Hanreich, R., Jalett, H. P., & Jelsch, E. (1999). The chiral switch of metolachlor: The development of a large-scale enantioselective catalytic process. *Chimia,* Vol.53, No.6, pp.275-280

Camilleri, P., Gray, A., Weaver, K., Bowyer, J. R. & Williams, D. J. (1989). Herbicidal diphenyl ethers stereochemical studies using enantiomers of a novel diphenyl ether phthalide. herbicidal diphenyl ethers stereochemical studies using enantiomers of a novel diphenyl ether phthalide. *J Agric Food Chem,* Vol.37, No.2, pp.519-523

Cai, X. Y., Liu, W. P., & Sheng, G. Y. (2008). Enantioselective degradation and ecotoxicity of the chiral herbicide diclofop in three freshwater alga cultures. *J Agric Food Chem,* Vol.56, No.6, pp.2139-2146

Chesters, G., Simsiman, G. V., Levy, J., Alhajjar, B. J., Fathulla, R. N., & Harkin, J. M. (1989). Environmental fate of alachlor and metolachlor. *Rev Environ Contamin Toxicol,* Vol.110, pp.1-74.

Chin, T. E., Wong, R. B., Pont, J. L., & Karu, A. E. (2002). Haptens and monoclonal antibodies for immunoassay of imidazolinone herbicides. *J Agric Food Chem,* Vol.50, No.12, pp.3380-3389

Couderchet, M., Bocion, P. F., Chollet, R., Seckinger, K., & Boger, P. (1997). Biological activity of two stereoisomers of the N-thienyl chloroacetamide herbicide dimethenamid. *Pesticide Science,* Vol.50, No.3, pp.221-227

Diao, J., Lv, C., Wang, X., Dang, Z., Zhu, W., & Zhou, Z.(2009). Influence of soil properties on the enantioselective dissipation of the herbicide lactofen in soils. *J Agric Food Chem,* Vol.57, No.13, pp.5865-5871

Diao, J., Xu, P., Wang, P., Lu, D., Lu, Y., & Zhou, Z. (2010). Enantioselective degradation in sediment and aquatic toxicity to Daphnia magna of the herbicide lactofen enantiomers. *J Agric Food Chem,* Vol.58, No.4, pp.2439-2445

Desiderio, C., Polcaro, C.M., Padiglioni, P. & Fanali, S. (1997). Enantiomeric separation of acidic herbicides by capillary electrophoresis using vancomycin as chiral selector. *J Chromatogr A,* Vol.781, No.1997, pp.503-513

Easson, L. H., & Stedman, E. (1933). Studies on the relationship between chemical constitution and physiological action: Molecular dissymmetry and physiological activity. *Biochem J,* Vol.27, No.4, pp.1257-1266

Fayez, K. A., & Kristen, U. (1996). The influence of herbicides on the growth and proline content of primary roots and on the ultrastructure of root caps. *Environ Exp Botany,* Vol.36, No.1, pp.71-81

Garrison, A. W. (2006). Probing the enantioselectivity of chiral pesticides. *Environ Sci Technol,* Vol.40, No.1, pp.16-23.

Gassmann, E., Kuo, J. E., & Zare, R. N. (1985). Electrokinetic separation of chiral compounds. *Science,* Vol.230, No.4727, pp.813-814

Gotz, T., & Boger, P. (2004). Very-long-chain fatty acid synthase is inhibited by chloroacetamides. *Abstracts of Papers of the American Chemical Society,* Vol.227, pp.52

Hallahan, B. J., Camilleri, P., Smith, A., & Bowyer, J. R. (1992). Mode of action studies on a chiral diphenyl ether peroxidizing herbicide: correlation between differential Inhibition of protoporphyrinogen IX oxidase activity and induction of tetrapyrrole accumulation by the enantiomers. *Plant Physiol,* Vol.100, No.3, pp.1211-1216

Hegeman, W. J., & Laane, R. W. (2002). Enantiomeric enrichment of chiral pesticides in the environment. *Rev Environ Contam Toxicol,* Vol.173, pp.85-116.

Imai, K., Kojima, H., Numata, T., Omokawa, H. & Tanaka, H.(2009). Chiral effects of (R)-/(S)-1-(α-methylbenzyl)-3-(p-tolyl)urea on the free amino acid levels in the root tips of rice and wheat. *Weed Biology and Management,* vol.9, pp.87–92

Jurado, A. S., Fernandes, M. A. S., Videira, R. A., Peixoto, F. P. & Vicente, J. A. F.(2011). Herbicides: the face and the reverse of the coin. An *in vitro* approach to the toxicity of herbicides in non-target organisms. *Herbicides and Environment,* InTech, ISBN 978-953-307-476-4, Rijeka, Croatia

Kazuhiro I., Hisahiro K., Takako N., Hiroyoshi O. & Hideyuki T.(2009). Chiral effects of (R)-/(S)-1-(α-methylbenzyl)-3-(p-tolyl)ureas on the free amino acid levels in the root tips of rice and wheat. Chiral effects of (R)-/(S)-1-(α-methylbenzyl)-3-(p-tolyl)ureas on the free amino acid levels in the root tips of rice and wheat. *Weed Biol Manag,* Vol.9, pp. 87–92

Kennard, C., Smith, G., & White, A. H. (1982). Structural aspects of phenoxyalkanoic acids-the structures of phenoxyacetic acid, (+/-)-2-phenoxypropionic acid, (+/-)-2-(4-chlorophenoxy)propionic acid, 2-methyl-2-phenoxypropionic acid and 2-(4-chlorophenoxy)-2-methylpropionic acid. *Acta Crystallographica Section B-Structural Science,* Vol.38, pp.868-875.

Koeller, K. M., & Wong, C. H. (2001). Enzymes for chemical synthesis. *Nature,* Vol.409, No.6817, pp.232-240.

Kondru, R. K., Wipf, P., & Beratan, D. N. (1998). Atomic contributions to the optical rotation angle as a quantitative probe of molecular chirality. *Science,* Vol.282, No.5397, pp.2247-2250

Kurihara, N., Miyamoto, J., Paulson, G. D., Zeeh, B., Skidmore, M. W., & Hollingworth, R. M. (1997). Pesticides report .37. Chirality in synthetic agrochemicals: Bioactivity and safety consideration (Technical report). *Pure Applied Chem,* Vol.69, No.9, pp.1335

Kunimitu, N., Masaaki, U., Takashi, I., Hajime, I. & Koichi S. (1988). Effect of (R)(+)-and(S)-(-)-quizalofop-ethyl on lipid metabolism in excised corn stem-base meristems. *J Pesticide Sci,* Vol.13, pp.269-276

Lao, W. J., & Gan, J. (2006). Responses of enantioselective characteristics of imidazolinone herbicides and Chiralcel OJ column to temperature variations. *J Chromatogr A,* Vol.1131, No.1-2, pp.74-84

Lewis, D. L., Garrison, A. W., Wommack, K. E., Whittemore, A., Steudler, P., & Melillo, J. (1999). Influence of environmental changes on degradation of chiral pollutants in soils. *Nature,* Vol.401, No.6756, pp.898-901

Li, H., Yuan, Y. L., Shen, C. S., Wen, Y. Z., & Liu, H. J. (2008). Enantioselectivity in toxicity and degradation of dichlorprop-methyl in algal cultures. *J Environ Sci Health B,* Vol.43, No.4, pp.288-292

Li, L., Zhou, S. S., Zhao, M. R., Zhang, A. P., Peng, H., & Tan, X., et al. (2008). Separation and aquatic toxicity of enantiomers of 1-(substituted phenoxyacetoxy)alkylphosphonate herbicides. *Chirality,* Vol.20, No.2, pp.130-138

Liu, H. J., Ye, W. H., Zhan, X. M., & Liu, W. P. (2006). A comparative study of rac- and S-metolachlor toxicity to Daphnia magna. *Ecotoxicol Environ Safety,* Vol.63, No.3, pp.451-455

Liu, H. J., & Xiong, M. (2009). Comparative toxicity of racemic metolachlor and S-metolachlor to Chlorella pyrenoidosa. *Aquat Toxicol,* Vol.93, No.2-3, pp.100-106

Liu, W. P., Ye, J., & Jin, M. Q. (2009). Enantioselective Phytoeffects of Chiral Pesticides. *Journal of Agricultural and Food Chemistry,* Vol.57, No.6, pp.2087-2095

Liu, W. P., Ye, J., & Jin, M. Q. (2009). Enantioselective phytoeffects of chiral pesticides. *J Agric Food Chem,* Vol.57, No.6, pp.2087-2095

Matell, M. (1953). Stereochemical studies on plant growth regulators. VII. Optically active α-(2-methyl-4-chlorophenoxy)-propionic acid and α-(2,4-dichlorophenoxy)-n-butyric acid and their steric relations. *Ark Kemi* Vol.6, pp.365-373

Muller, B. P., Zumdick, A., Schuphan, I., & Schmidt, B. (2001). Metabolism of the herbicide glufosinate-ammonium in plant cell cultures of transgenic (rhizomania-resistant) and non-transgenic sugarbeet (Beta vulgaris), carrot (Daucus carota), purple foxglove (Digitalis purpurea) and thorn apple (Datura stramonium). *Pest Manag Sci,* Vol.57, No.1, pp.46-56

Muller, T. A., Fleischmann, T., van der Meer, J. R., & Kohler, H. P. (2006). Purification and characterization of two enantioselective alpha-ketoglutarate-dependent dioxygenases, RdpA and SdpA, from Sphingomonas herbicidovorans MH. *Appl Environ Microbiol*, Vol.72, No.7, pp.4853-4861

Muller, T. A., & Kohler, H. P. (2004). Chirality of pollutants--effects on metabolism and fate. *Appl Microbiol Biotechnol*, Vol.64, No.3, pp.300-316

Omokawa, H., Ryoo, J. H., & Kashiwabara, S. (1999). Enantioselective relieving activity of methylbenzylphenylureas toward bensulfuron-methyl injury to rice. enantioselective relieving activity of methylbenzylphenylureas toward bensulfuron-methyl injury to rice. *Biosci Biotechnol Biochem*, Vol.63, No.2, pp.349-355

Omokawa, H. & Ryoo, J. H. (2001). Enantioselective Response of Rice and Barnyard Millet on Root Growth Inhibition by Optically Active α-Methylbenzyl Phenylureas. Enantioselective Response of Rice and Barnyard Millet on Root Growth Inhibition by Optically Active α-Methylbenzyl Phenylureas. *Pestic Biochem Physiol*, Vol.70, No.1, pp.1-6

Omokawa, H., Murata, H., & Kobayashi, S. (2004). Chiral response of Oryzeae and Paniceae plants in alpha-methylbenzyl-3-p-tolyl urea agar medium. *Pest Manag Sci*, Vol.60, No.1, pp.59-64

Pirkle, W. H., Lee, W., & Welch, C. J. (1997). Chromatographic separation of the enantiomers of 2-aryloxypropionic acids, esters and amides. *Enantiomer*, Vol.2, No.6, pp.423-431

Polcaro, C. M., Berti, A., Mannina, L., Marra, C., Sinibaldi, M., & Viel, S. (2004). Chiral HPLC resolution of neutral pesticides. *J Liquid Chromatogr Related Technol*, Vol.27, No.1, pp.49-61

Qian, H. F., Hu, H., Mao, Y., Ma, J., Zhang, A., & Liu, W. P., et al. (2009). Enantioselective phytotoxicity of the herbicide imazethapyr in rice. *Chemosphere*, Vol.76, No.7, pp.885-892

Qian, H. F., Lu, T., Peng, X., Han, X., Fu, Z. W., & Liu, W. P. (2011). Enantioselective phytotoxicity of the herbicide imazethapyr on the response of the antioxidant system and starch metabolism in arabidopsis thaliana. *PLoS One*, Vol.6, No.5, pp.e19451

Roelfes, G. (2007). DNA and RNA induced enantioselectivity in chemical synthesis. *Mol Biosyst*, Vol.3, No.2, pp.126-135

Ruhland, M., Engelhardt, G., & Pawlizki, K. (2002). A comparative investigation of the metabolism of the herbicide glufosinate in cell cultures of transgenic glufosinate-resistant and non-transgenic oilseed rape (Brassica napus) and corn (Zea mays). *Environ Biosafety Res*, Vol.1, No.1, pp.29-37

Ryoo, J. H., Kuramochi, H. & Omokawa, H. (1998). Enantioselective herbicidal activity of chiral α-methylbenzylphenylureas against Cyperaceae and Echinochloa paddy weeds. Enantioselective herbicidal activity of chiral α-methylbenzylphenylureas against Cyperaceae and Echinochloa paddy weeds. *Biosci Biotechnol Biochem*, Vol.62, No.11, pp.2189-2193

Sakata, G., Makino, K., Kusano, K., Satow, J., Ikai, T., & Suzuki, K. (1985). Preparation of optically pure ethyl (r)-(+) and (s)-(-)-2-[4-(6-chloro-2-quinoxalinyloxy)phenoxy]propanoate by resolution method and their herbicidal activities. *J Pestic Sci*, Vol.10, No.1, pp.75-79

Schmalfuss, J., Matthes, B., Knuth, K., & Boger, P. (2000). Inhibition of acyl-CoA elongation by chloroacetamide herbicides in microsomes from leek seedlings. *Pestic Biochem Physiol*, Vol.67, No.1, pp.25-35

Schneiderheinze, J. M., Armstrong, D. W., & Berthod, A. (1999). Plant and soil enantioselective biodegradation of racemic phenoxyalkanoic herbicides. *Chirality*, Vol.11, No.4, pp.330-337

Sekhon, B. S. (2009). Chiral pesticides. *J Pestic Sci*, Vol.34, No.1, pp.1-12

Shimabukuro, R. H., & Hoffer, B. L. (1995). Enantiomers of diclofop-methyl and their role in herbicide mechanism of action. *Pestic Biochem Physiol*, Vol.51, No.1, pp.68-82

Spangler, L. A., Mikolajczyk, M., Burdge, E. L., Kielbasinski, P., Smith, H. C., & Lyzwa, P., et al. (1999). Synthesis and biological activity of enantiomeric pairs of phosphosulfonate herbicides. *J Agric Food Chem*, Vol.47, No.1, pp.318-321

Stuart, D. A., & Mccall, C. M. (1992). Induction of somatic embryogenesis using side-chain and ring modified forms of phenoxy acid growth-regulators. *Plant Physiol*, Vol.99, No.1, pp.111-118

Tan, S., Evans, R., & Singh, B. (2006). Herbicidal inhibitors of amino acid biosynthesis and herbicide-tolerant crops. *Amino Acids*, Vol.30, No.2, pp.195-204

Wen, Y. Z., Yuan, Y. L., Shen, C. S., Liu, H. J., & Liu, W. P. (2009). Spectroscopic investigations of the chiral interactions between lipase and the herbicide dichlorprop. *Chirality*, Vol.21, No.3, pp.396-401

Wen, Y. Z., Chen, H., Yuan, Y. L., Xu, D. M., & Kang, X. (2011). Enantioselective ecotoxicity of the herbicide dichlorprop and complexes formed with chitosan in two fresh water green algae. *J Environ Monit*, Vol.13, No.4, pp.879-885

Wen, Y. Z., Yuan, Y. L., Chen, H., Xu, D. M., Lin, K. D., & Liu, W. P. (2010). Effect of chitosan on the enantioselective bioavailability of the herbicide dichlorprop to Chlorella pyrenoidosa. *Environ Sci Technol*, Vol.44, No.13, pp.4981-4987

Whittington, J., Jacobsen, S. & Rose, A. (2001). Optically pure clethodim, compositions and methods for controlling plants. Optically pure clethodim, compositions and methods for controlling plants. *United States Patent*

Wong, C. S. (2006). Environmental fate processes and biochemical transformations of chiral emerging organic pollutants. *Anal Bioanal Chem*, Vol.386, No.3, pp.544-558

Xu, D. M., Wen, Y. Z. & Wang, K. (2010). Effect of chiral differences of metolachlor and its (S)-isomer on their toxicity to earthworms. *Ecotoxicol Environ Safety* ,Vol.73, No.2010, pp.1925-1931

Xu, D. M., Xu, X., & Liu, W. P. (2009). Acute toxicity differece of metolachlor and its S-isomer on earthworms. *J Environ Sci-China*, Vol.29, No.9, pp. 1000-1004

Ye, J., Zhang, Q., Zhang, A. P., Wen, Y. Z., & Liu, W. P. (2009). Enantioselective effects of chiral herbicide diclofop acid on rice Xiushui 63 seedlings. *Bull Environ Contam Toxicol*, Vol.83, No.1, pp.85-91

Yoon, T. P., & Jacobsen, E. N. (2003). Privileged chiral catalysts. *Science*, Vol.299, No.5613, pp.1691-1693

Zhan, X. M, Liu, H. J., Miao, Y. G., Liu, W. P. (2006). A comparative study of rac- and S-metolachlor on some activities and metabolism of silkworm, *Bombyx mori* L. *Pestic Biochem Physiol*, Vol.85, No.2006, pp.133–138

Zhou, Q. Y., Liu, W. P., Zhang, Y. S., & Liu, K. K. (2007). Action mechanisms of acetolactate synthase-inhibiting herbicides. *Pestic Biochem Physiol*, Vol.89, No.2, pp.89-96

Zhou, Q. Y., Xu, C., Zhang, Y. S., & Liu, W. P. (2009). Enantioselectivity in the phytotoxicity of herbicide imazethapyr. *J Agric Food Chem*, Vol.57, No.4, pp.1624-1631

Zipper, C., Nickel, K., Angst, W., & Kohler, H. P. (1996). Complete microbial degradation of both enantiomers of the chiral herbicide mecoprop [(RS)-2-(4-chloro-2-methylphenoxy)propionic acid] in an enantioselective manner by Sphingomonas herbicidovorans sp. nov. *Appl Environ Microbiol*, Vol.62, No.12, pp.4318-4322

Part 2

Mode of Action

5

Weed Resistance to Herbicides in the Czech Republic: History, Occurrence, Detection and Management

Kateřina Hamouzová[1*], Jaroslav Salava[2], Josef Soukup[1],
Daniela Chodová[2] and Pavlína Košnarová[1]
*[1]Czech University of Life Sciences Prague,
Department of Agroecology and Biometerology,
Kamýcká, Prague 6-Suchdol,
[2]Crop Research Institute, Division of Plant Health, Prague 6-Ruzyně,
Czech Republic*

1. Introduction

Herbicide resistance results from the repeated selection for resistant individuals in field populations. Herbicide resistance has evolved to many important groups of herbicides, encompassing almost all of the major herbicide mode-of-action groups (Heap, 2000).

In the Czech Republic, herbicide resistance has been reported in 16 weed species, mostly found on non-arable land (railways) and in orchards with a history of intensive herbicide use. In maize and sugar beet, *Echinochloa cruss-galli, Chenopodium album, Chenopodium rigidum* and *Solanum nigrum* have been identified with resistance to photosystem 2 inhibitors. Seven *Kochia scoparia* biotypes resistant to chlorsulfuron have been found on railways, roadsides and car parks in the Czech Republic since 1996 (Salava *et al.*, 2004). *Alopecurus myosuroides* is the weed with local importance in the Czech Republic. However, the poor weed control had been reported in several fields and the tested populations showed sulfonylurea resistant phenotype (Slavíková *et al.*, 2011). On arable land, the most important seems to be ALS-resistant *Apera spica-venti* (Hamouzová *et al.*, 2011). ALS-resistant *A. spica-venti* was first documented in 2005, in populations collected from fields in western and southern part of the Czech Republic that had continuous chlorsulfuron treatment for at least 5 years (Hamouzová *et al.*, 2011). To date *A. spica-venti* resistant to ALS herbicides has been reported in several countries in Europe (Heap, 2011).

In this chapter, history and present state of weed resistance to herbicides in the Czech Republic is reviewed, with emphasis on origin, development, level of resistance, methods of testing for herbicide resistance, mechanisms, molecular basis, and effective management of herbicide resistance.

---

* Corresponding Author

2. Weed resistance to photosystem 2-inhibiting herbicides

Triazine herbicides were repeatedly applied on railways in the Czech Republic for about 20 years. The use of atrazine herbicides was terminated in 1992 as a consequence of their long persistence or tendency to move off site, damage of surrounding vegetation, enter watercourses and detection of resistant biotypes of ten weed species (Anonymous, 2010). Atrazines were replaced with terbuthylazine. Meanwhile, plants of *Digitaria sanquinalis* that are not susceptible to terbuthylazine are found on the Czech railways.

A biotype *of Digitaria sanquinalis* resistant to atrazine was first described by Gasquez in 1983 (Heap, 2011). Resistance to triazine herbicides can be conferred by a modified target site (mutation in the *psb*A gene which codes for the D1 protein of PS2) or an enhanced detoxification of herbicide (Robinson and Greene, 1976; Gawronski *et al.*, 1992; Monteiro and Rocha, 1992; De Prado *et al.*, 1999 and Chodová *et al.*, 2004).

3. Materials and methods

Seed source

Seeds of *Digitaria sanquinalis* were originally collected from plants growing on the railway tracks and in the vicinity of the railway station Prague-Libeň in 2005. Plants were grown in a greenhouse under controlled conditions. The presence of individuals resistant to atrazine in the population was confirmed by the whole-plant response assay. Seeds of susceptible biotypes were collected from the fields, private gardens or places without application of herbicides.

Whole-plant response assay

Seeds were sown into plastic pots 10x10 cm in size, 0.2 g seeds per pot, and plants were thinned to 5-10 individuals per pot after their emergence. The pots were placed in a greenhouse (temperature 20-24°C, relative air humidity 70-90%, no supplemental lighting).

Herbicides were applied at the stage of 2-4 right leaves of plants with a hand sprayer using 2 000 g.ha[-1] of atrazine and simazine (Gesaprim 500 g a.i. l[-1], Novartis Crop Protection AB, Basel, Switzerland; Gesatop 900 g a.i. kg[-1], Ciba-Geigy AB, Basel, Switzerland). The effect of the herbicide was assessed 15 days after treatment. Phytotoxic symptoms were ranked individually according to the European Weed Research Society classification scheme for plant tolerance (Frey *et al.*, 1999).

Chlorophyll fluorescence parameters

Chlorophyll fluorescence for *Digitaria sanquinalis* was measured in 10 plants using a chlorophyll fluorometer FMS2 (Hansatech Instruments, UK). The leaf discs were cut from the middle part of leaf blade and soaked at 1, 2, 3 and 4 hour-intervals in either 1% atrazine solution or water. Measurements were done on 30-min dark-adapted leaves. Maximum quantum efficiency of PS 2 photochemistry, i.e. F_v/F_m [defined as $[(F_m - F_o)/F_m]$, where F_o is the minimum, and F_m is the maximum chlorophyll fluorescence yield in the dark-adapted state], was calculated.

Biotype	Hours after treatment	F_0		F_m		F_v/F_m		Differences	
		1% atrazine	water	1% atrazine	water	1% atrazine	water	A	B
Susceptible	1	383	241	1107	1200	0.654	0.799	*	*
	2	391	239	1247	1243	0.686	0.808	*	*
	3	323	242	1055	1250	0.694	0.806	*	*
	4	494	241	939	1220	0.474	0.802	**	**
Resistant	1	341	267	1099	1118	0.690	0.761	**	*
	2	311	295	1101	1253	0.718	0.765	*	*
	3	328	281	1165	1152	0.718	0.756	ND	*
	4	313	313	1147	1159	0.727	0.730	ND	**

A - *significant differences at 0.05 probability level, **significant differences at 0.01 probability level in F_v/F_m values of susceptible (or resistant) plants treated with atrazine compared with relevant control sample (water), NDsignificance of difference not determined
B - *significant differences at 0.05 probability level, **significant difference at 0.01 probability level in F_v/F_m values between resistant and susceptible biotype after 1 to 4 hours exposure to the herbicide

Table 1. Chlorophyll fluorescence measurements on dark-adapted leaves of *Digitaria sanquinalis*.

Photochemical activity of isolated chloroplasts

Isolation of chloroplasts from leaves and measurement of their activity was performed as described in Holá et al. (1999). Photochemical activity was determined polarographically as the Hill reaction activity (HRA), characterized as the amount of oxygen formed by the chloroplast suspensions in defined conditions after white light irradiation and addition of an artificial electron acceptor.

Atrazine concentration	Susceptible biotype	Resistant biotype
0 M	46.07 ± 4.77	51.22 ± 2.30
0.0000001 M	37.34 ± 4.83	47.06 ± 3.73
0.000001 M	6.55 ± 0.73**	43.37 ± 0.92
0.00001 M	0 ND	30.92 ± 1.39*
0.0001 M	0 ND	22.61 ± 0.46*
0.001 M	0 ND	7.39 ± 0.93**

*difference (compared to the relevant sample measured without the addition of atrazine) significant at 0.05 probability level, ** difference significant at 0.01 probability level, ND significance of difference not determined

Table 2. Hill reaction activity (mean values in mmol O_2 kg^{-1} of chlorophyll s^{-1} ± standard error) of mean in suspensions of chloroplasts isolated from leaves of four atrazine -resistant or -susceptible plants of *Digitaria sanguinalis* after the addition of various concentrations of atrazine.

Gene sequencing

Plants of which susceptibility or resistance was confirmed using the whole-plant response assay (for details see Nováková et al., 2005), Hill reaction activity and chlorophyll fluorescence assay were chosen for molecular analysis. DNA of individual plants was extracted using the DNeasy Plant Mini Kit (QIAGEN, Hilden, Germany) following the manufacturer's instructions. Polymerase chain reaction (PCR) and sequencing of the *psb*A gene were conducted using standard methods as described previously (Nováková *et al.* 2005). Sequence editing and analysis were done using the BLAST program (Altschul *et al.* 1997). The ExPASy translate tool (Gasteiger *et al.* 2003) was used to determine the peptide sequences.

```
                                         264
R                                        Gly
S     Gly Arg Leu Ile Phe Gln Tyr Ala Ser Phe Asn Asn Ser Arg Ser Leu His Phe
S     GGT CGA TTA ATC TTC CAA TAT GCT AGT TTC AAC AAC TCT CGT TCT TTA CAC TTC
R     ..................................G..................................
```

Fig. 1. Alignment of cpDNA encompassing the atrazine resistance-conferring region of the *psb*A gene of *Digitaria sanguinalis* using susceptible (S) as a reference. Dots in the resistant (R) sequence indicate matches to the reference sequence; differences are indicated by A, C, G or T. Bold print in the amino acid sequence indicates the site where the mutation confers atrazine resistance.

4. Results

The screening of large crabgrass plants obtained from the railway station Prague-Libeň revealed the presence of individual plants resistant to atrazine in this population. The resistance to atrazine was confirmed by the *in vivo* measurement of chlorophyll fluorescence (Table 1) emitted by leaves treated with this active ingredient, as well as by the measurements of the Hill reaction activity in suspensions of isolated chloroplasts (Table 2) after the addition of atrazine solution.

After atrazine treatment, leaves from the resistant biotype did not show any changes in the fluorescence curve pattern when compared with the control leaves treated with water. On the other hand, leaves from the susceptible biotype showed changed fluorescence and the F_v/F_m ration was considerably reduced.

The Hill reaction activity in suspensions of chloroplasts isolated from large crabgrass plants susceptible to atrazine distinctively decreased with the addition of atrazine to these suspensions and its values were zero after the treatment with 0.01 mM (or higher) atrazine solutions. Contrary to this, chloroplasts isolated from leaves of the resistant biotype showed certain Hill reaction activity even with the 1 mM atrazine treatment and the decline with the increasing atrazine concentration was not as pronounced as in the previous case (Table 2). The ED_{50} value for the susceptible biotype was 0.17 µM, for the resistant biotype 29.21 µM, and RF_{50} was calculated as 166.

A region of the chloroplast *psb*A gene that encodes for amino acids 163 to 329 of the D1 protein of PS2 was selectively amplified using PCR. Sequence analysis of the fragment from the herbicide-resistant biotype of large crabgrass exhibited a substitution from serine to glycine at position 264 of the D1 protein of PS2 (Figure l).

5. Discussion

Recent screening of large crabgrass in the Czech Republic (particularly along railways and other transport corridors) has revealed the presence of atrazine-resistant plants of this weed species that tolerated the dose of 3 000 g atrazine.ha[-1] at the railway station Prague-Libeň. Since the active ingredient atrazine was irregularly used along the Czech railways only till 1992 (and again for a short period 1997-1999), these resistant individuals had to grow up from seeds that retained their germinability in the specific railway conditions for fairly long time.

Besides the basic screening tests commonly used for the determination of plant susceptibility or resistance to atrazine, evaluation of photosynthetic parameters can be with advantage used for the evaluation of resistance in individual plants. The mutation in the 264th codon of the *psbA* gene, usually found in atrazine-resistant plants (Hischberg and McIntosh, 1983) reduces binding capacity of triazine herbicides to the D1 protein, thus enabling the electron transfer in PS2 even in the presence of atrazine. Our measurements of chlorophyll fluorescence and/or photochemical activity of isolated chloroplasts confirmed this, as we found that PS2 in atrazine-resistant plants of large crabgrass from the railway station Prague-Libeň indeed retains its activity after treatment of leaf discs or chloroplasts with atrazine, though to a lesser degree than in the untreated plants. There were some differences in chlorophyll fluorescence curves as well, similarly to the findings of De Prado *et al.* (1999).

The photochemical activity of chloroplasts isolated from atrazine-resistant plants, measured without the addition of atrazine to chloroplast suspensions, did not significantly differ from the activity of chloroplasts isolated from atrazine-susceptible plants, contrary to the previous findings for other weed species (e.g. Ort *et al.*, 1983; Sundby *et al.*, 1993; Chodová *et al.*, 1994; Körnerová *et al.*, 1998; Devine and Shukla, 2000). This was rather interesting, as the mutation in D1 protein usually decreases the efficiency of electron transport in PS2 under normal conditions. Gadamski *et al.* (1996), who characterized atrazine-resistant biotypes of large crabgrass in several Polish populations, did not find the target-site mutation at codon 264 of the *psbA* gene and suggested the possibility of different mechanism of resistance (*e.g.* non-target site). We have analysed the herbicide-binding region of the *psbA* gene and found out that in our case the molecular basis of resistance to atrazine in large crabgrass is the „classical" G264S substitution, similarly to other atrazine-resistant biotypes of various weed species (*e.g.* *Kochia scoparia, Solanum nigrum, Senecio vulgaris*) found in the Czech Republic previously (Chodová and Salava, 2004; Salava *et al.*, 2004; Nováková *et al.*, 2005). Moreover, there was an excellent correspondence between the presence of the mutation and phenotypic resistance to atrazine of individual plants, as determined by the assays of chlorophyll fluorescence or photochemical activity of chloroplasts. It thus seems that the Czech biotype of atrazine-resistant large crabgrass is not related to the Polish ones.

6. Weed resistance to acetolactate synthase-inhibiting herbicides

Acetolactatesynthase (ALS), also known as acetohydroxyacid synthase (AHAS) is required for the production of amino acids valine, leucine and isoleucine (Duggleby and Pang, 2000). The branched-chain amino acids are synthesized by plants, algae, fungi, bacteria and archaea, but not by animals, therefore are potential targets for the development of herbicides (McCourt and Duggleby, 2006). Worldwide, herbicides inhibiting acetolactatesynthase (ALS) are the most used ones at present. ALS is the primary target site of action for five structurally distinct classes of herbicides including pyrimidinylthiobenzoates, sulfonylureas, imidazolinones, triazolopyrimidine sulfonamides and sulfonylaminocarbonyltriazolinones (Shimizu *et al.*, 2002).

In 1987, *Lactuca serriola* was the first resistant weed species reported to be selected by ALS-inhibiting herbicides (Mallory-Smith *et al.*, 1990).

Since then, rapid emergence of resistance to ALS-inhibitors has been identified in 110 weed species around the world (Heap, 2011).

Currently, five mechanisms of herbicide resistance have been identified in weeds: 1) altered target site due to a mutation at the site of herbicide action, 2) metabolic deactivation the active ingredient is transformed to nonphytotoxic metabolites, 3) reduce absorption and/or translocation of herbicide, 4) sequestration/compartmentation herbicide is located in vacuoles or cell walls and 5) gene amplification/over-expression of the target site, respectively (Nandula *et al.*, 2010).

In Europe, to the most important herbicide-resistant weeds belong blackgrass (*Alopecurus myosuroides*), *Lolium* species, *Papaver rhoeas* and wild oat (*Avena* spp.) (Moss *et al.*, 2007). As seen from the list of resistant species, the grass weeds are on the top of farmers and scientific concern.

The occurrence of grass weeds is strongly influenced by geographical factors. In the areas with predominate ocean climate (such as the Great Britain, north France, western part of Germany, Denmark), the most harmful grass weed is *Alopecurus myosuroides* and *Lolium multiflorum*, in the Mediterranean Europe to the most detrimental weed belongs *Avena fatua*, *Avena sterilis* and *Lolium rigidum*. In the central and north-west Europe, *Apera spica-venti* is the biggest threat; *Bromus sterilis* and *Avena fatua* propagate quickly in the recent years. Generally, the protection against certain weed grasses in cereals is expensive and the spectrum of herbicides is not broad. Interaction of cultural methods and plant protection methods supports the occurrence of weeds and in many cases the herbicide-resistance. At present, populations with resistance to herbicides with different modes of action can be found across the world, whilst the information on extent of herbicide-resistant species has been best reported in the North America and Europe (Moss, 2002).

7. Reasons of origin of resistance in *Kochia scoparia* and *Apera spica-venti*

Crop rotation dominated by winter cereals and often autumn sown crops, together with reduced soil tillage, favour the spread of *A. spica-venti*, encouraging its life cycle (winter annual weed) reproducing exclusively generatively (Soukup *et al.*, 2006). In recent years, *A. spica-venti* is progressively occurring in spring cereals.

Repetitive use of the same mode of action herbicides resulted in the frequent appearance of ALS-resistant *A. spica-venti*. It was proven that ALS-inhibitors are the most resistance-prone herbicide group (Délye and Boucansaud, 2008) and has occurred with only three applications of these herbicides (Preston *et al.*, 1999). Major factors influence the evolution of herbicide resistance including the intensity of selection by herbicide and the initial frequency of herbicide resistant individuals in the populations (Preston and Powles, 2002).

8. Material and methods for identification of resistant weeds

Reliable diagnostic testing methods for determining weed population on resistance to ALS inhibitor are an important aspect. Herbicide resistance can be studied by different methods, their selection depends on the purpose of the scientific research (confirmation and characterization of resistance, examination of the biological and physiological traits of resistance, weed management, etc.). Various methods have been developed for confirming ALS resistance in grass weeds, these ranged from field experiments over whole-plant pot trials carried out under controlled conditions. To date, the whole-plant test remains the most commonly employed method for confirming resistance (Beckie *et al.*, 2000). However, there are many aspects significantly affecting the response to herbicides in whole-plant assays. Among the most important variables are formulation, adjuvant, spray volume, time of application as well as growth stage.

In those biotypes, where resistance was confirmed, more sophisticated laboratory based enzyme and DNA/RNA assays have been used and optimized for *A. spica-venti* (Hamouzová *et al.*, 2011; Massa *et al.*, 2011). For the quick detection of amino acid substitutions cleaved amplified polymorphic sequences markers can be used and partial *als* gene sequencing (Kaundun and Windass, 2006; Délye and Boucansaud, 2008). Most applicable to testing for ALS resistance are quantitative/RT-PCR and SNP analysis. Most recently, pyrosequencing technology is an established method for genotyping of SNPs, enabling rapid real-time sequence determination (Wagner *et al.*, 2008). The specificity of molecular techniques is a major advantage for research purposes. There is a need for the routine resistance screening which will be cheap and available for the general public. One of the approaches could be Syngenta RISQ method based on screening the survivorship of the seedlings in Petri dishes exposed to the different doses of herbicides (Kaundun *et al.*, 2011).

To assess the mechanism of resistance the HPLC methods (Chodová *et al.*, 2009; Massa *et al.*, 2011) and uptake and translocation of radiolabelled herbicides (Everman *et al.*, 2009) are used.

The interpretation of the results cannot be missed and is dependent on methods used, the mechanism of resistance, and proportion of plants within the population. The comparison of results deriving from different methods is not straightforward as the data on genetic, enzyme level and whole plant level cannot be compared e.g., herbicide efficacy under controlled conditions is usually significantly higher than in fields. The proper statistical evaluation of the data obtained from different datasets is crucial to make a right decision if the population is/is not resistant to herbicide.

Seed sources

The susceptible population of *A. spica-venti* had never been exposed to ALS-inhibiting herbicides and was sampled on the organic farm in the Czech Republic. The other susceptible population was obtained from Syngenta (Hamouzová *et al.*, 2011).

Seeds of *A. spica-venti* plants surviving applications of the registered field dose of ALS inhibiting herbicides were collected from winter wheat fields in July 2004-2010. Seed samples were obtained from fields with poor weed control and were not derived from the random sampling, thus extrapolation of results was not possible. Site selection was based on farmers' suspicion and past herbicide use. There were tested more than 100 populations using 12 different herbicides (Table 3) in total.

	Chlorsulfruon (15g a.i. ha⁻¹)	Sulfosulfuron (19.5g a.i. ha⁻¹)	Iodosulfuron (7.5g a.i. ha⁻¹)	Mesosulfuron-methyl (4.5g a.i. ha⁻¹) + Iodosulfuron (0.9g a.i. ha⁻¹)	Sulfometuron methyl (52.5g a.i. ha⁻¹)	Pinoxaden (30g a.i. ha⁻¹)
Number of samples tested	158	40	15	16	15	53
Samples showing resistance	118	22	13	16	13	0

	Isoproturon (1000g a.i. ha⁻¹)	Fenoxaprop –P- ethyl (69g a.i. ha⁻¹)	Prosulfocarb (1600g a.i. ha⁻¹)	Propoxycarbazone (42g a.i. ha⁻¹)	Pyroxsulam (18.75g a.i. ha⁻¹)	Imazetapyr-NH₄ (50g a.i. ha⁻¹)
Number of samples tested	72	26	6	5	70	3
Samples showing resistance	4	0	0	5	55	1

Note: In the table, populations in which herbicide efficacy was lower than 50% as compared to untreated are mentioned.

Table 3. Overview of *A. spica-venti* populations tested in whole-plant assays to herbicides with different mode of action at recommended doses.

Whole-plant response assay

Whole-plant response assays are the most widely used methods, while mimic the real field conditions the best. To characterise the susceptibility of individual populations of *A. spica-venti* to herbicides and calculate the resistance factor (RF), dose-response assays were performed.

Seed were sown in plastic pots (25 cm²) filled with a chernozem soil [clay content 46% (loamy soil), soil pH (KCl) 7.5, sorption capacity of soil: 209 mmol (+), 87 mg kg⁻¹ P, 203 mg kg⁻¹ K, 197 mg kg⁻¹ Mg, 8073 mg kg⁻¹ Ca] and placed in a greenhouse. For screening resistance to commercially formulated herbicides applied post-emergently, the herbicide was applied to grass weed at the two- to three-leaf stage using a laboratory chamber sprayer equipped with a nozzle Lurmark 015F80 delivering 250 l ha⁻¹. The recommended doses of herbicides were included and the doses ranged from 0.01 to 150-fold of recommended one were applied. The herbicide dose ranges for the S biotypes differed from those of R biotypes, to provide more accurate models. Pot experiments were conducted in a randomized block design containing four replicates. Herbicide susceptibility was assessed 4 weeks after treatment, when plants from the susceptible population were all clearly controlled. Visible symptoms on surviving plants were expressed as a percentage (0% no survival, 100% plants without any visible damage) of the untreated pots for the same population. Plants were cut at the soil surface, oven dried for 48 h at 105°C and weighed. Data were fitted to the non-linear regression using the Weibull four-parameter function where parameters are as follows: the dependent variable (efficacy or enzyme activity expressed as percentage of the untreated control, dry biomass weight in grams), the coefficients corresponding to the lower and upper limits, the herbicide dose at the point of inflection halfway between the upper and lower asymptotes (GR_{50}, I_{50}), and the slope around inflection point, independent variable is the herbicide dose (Nielsen et al., 2004; Ritz et al., 2006) The statistical software R 2.8.0 was used (R Development Core Team, 2008).

Figure 2 illustrates the resistance pattern as found in 3 A. spica-venti biotypes from the Czech Republic compared to susceptible standard. Growth reduction of 50% could not be achieved for some resistant biotypes even at very high application doses of some herbicides (more than 100-times higher than the recommended dose).

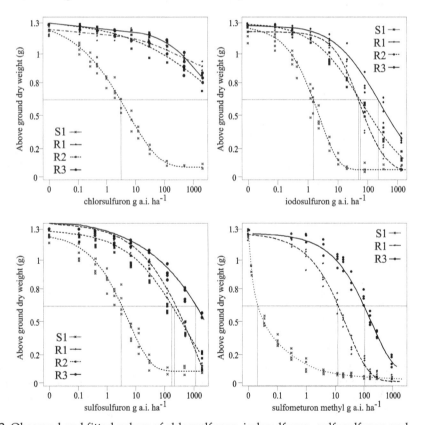

Fig. 2. Observed and fitted values of chlorsulfuron, iodosulfuron, sulfosulfuron and sulfometuron-methyl above ground dry biomass weight (g) with 3 resistant and 1 susceptible biotype of A. spica-venti (adapted from the paper of Hamouzová et al., 2011).

In vitro acetolactate synthase activity

The use of ALS *in vitro* assay is effective for determining the involvement of target-site insensitivity in conferring resistance to the herbicide studied. ALS activity was measured by estimation of the product, α-acetolactate, after its conversion to acetoin by decarboxylation in the presence of acid (Ray, 1984). The protein concentration was determined by the method of Bradford (1976) using bovine serum albumin as a standard. *A. spica-venti* seedlings were grown in pots. Seedlings were harvested 28 d after sowing when at the tree- to four-leaf stage. Two grams of leaf tissue from *A. spica-venti* were harvested and homogenised in liquid nitrogen and mixed with extraction buffer [5 mM $MgCl_2$, 50mM thiamine pyrophosphate (TPP), 1 mM flavine adenine dinucleotide (FAD), 10 mM sodium pyruvate and 126 g L^{-1} glycerol]. The mixture was homogenized during 10 min using magnetic stirrer. The homogenate was filtered through six layers of cheesecloth and centrifuged 10,000g for 15 min. The ALS was precipitated from the supernatant at 50% of $(NH_4)_2SO_4$ saturation. The mixture was stirred during 30 min after $(NH_4)_2SO_4$ addition and then centrifuged 20,000g for 20 min. The final pellet was resuspended in extraction buffer, and the enzyme was desalted on Sephadex G-25 PD-10 columns equilibrated with elution buffer (1 M potassium phosphate (KH_2PO_4/K_2HPO_4), pH 7.5; 50 mM sodium pyruvate; 0.1 M $MgCl_2$). Acetolactate synthase activity was assayed by adding enzyme extract to freshly prepared assay buffer (1 M KH_2PO_4/K_2H-PO_4), pH 7.5; 0.5 M sodium pyruvate; 0.1 M $MgCl_2$, 50 mM TPP; 1 mM FAD), with concentrations of herbicides increasing from 10^{-2} to 10^8 nM active ingredient. The assay mixture was incubated at 37°C for 2 h, and the reaction was terminated by the addition of 6 M H_2SO_4. Acetoin was detected as a coloured complex ALS activity was formed after addition of creatine and 1-naphtol and absorbance was measured at wavelength of 540 nm. Background was subtracted using control tubes where the reaction was stopped before the incubation. Specific enzyme activity was calculated from the standard curve and was expressed as μmol acetoin mg^{-1} protein h^{-1}. Data from experiment were fitted to the log–logistic model described above. The equations obtained were used to calculate the herbicide concentrations that reduced enzyme activity by 50% (I_{50}) of the untreated control. R/S ratios were calculated dividing the I_{50} of the R population by the I_{50} of the S population.

The data in Table 4 provide different I_{50} values for 5 putative resistant biotypes of *A. spica-venti*. Statistically significantly higher values of I_{50} proved that resistant plants have an altered form of the ALS enzyme. The level of resistance represented by RF varied significantly between the biotypes. The highest resistant ALS to chlorsulfuron was observed for the biotypes R3 and R4, while the biotypes R1 and R5 were less resistant. The isolation of ALS was similar among biotypes examined on the basis of the extractable protein, however all resistant biotypes had higher amounts of extracted leaf protein.

Molecular analysis of *ALS* gene in *Kochia scoparia* and *Apera spica-venti*

The modifications identified in naturally occurring plant populations occur in one of seven amino acids in the ALS enzyme where numbering was standardised in accordance with the *Arabidopsis thaliana* L. Heynh. sequence: Ala122, Pro197, Ala205, Asp376, Trp574, and Ser653 and Gly654 (Tranel and Wright, 2002; Powles and Yu, 2010). To date, twenty-two amino acid substitutions at all domains have been identified (Yu et al., 2010). Other mutation sites that give resistance have been created in the laboratory but are not known to occur naturally. The most commonly encountered mutations in weed grasses involve the residues of proline

Biotype	B	C	D	I_{50} (nM)	RF	ALS specific activity (μmol acetoin mg^{-1} protein h^{-1})	Protein concentration (μg protein mg^{-1} fresh weight)
S1	0.441	0	1	30.39	-	33.09	1.29 ± 0.13
S2	0.515	0	1	68.51	2.25	27.80	1.16 ± 0.19
R1	0.280	0	1	1318.23	43.37	38.34	1.45 ± 0.37
R2	0.366	0	1	3815.71	125.55	38.89	1.39 ± 0.22
R3	0.245	0	1	119514.82	3932.70	30.20	1.34 ± 0.18
R4	0.380	0	1	5160.17	169.79	38.95	1.46 ± 0.38
R5	0.420	0	1	1830.42	60.21	44.77	1.52 ± 0.20

S – susceptible standard, R1-R5 – resistant biotypes, B – slope of the curve around I_{50}, C – minimum ALS activity, D – maximum ALS activity, I_{50} – concentration of chlorsulfuron that causes 50 % inhibition of ALS activity related to the maximal activity reached, RF – resistance factor = I_{50} R/ I_{50} S,

Table 4. Parameters of the log-logistic dose-response model describing the ALS-activity in *Apera spica-venti* as affected by chlorsulfuron (adopted from the paper of Nováková *et al.*, 2006).

at position 197 and tryptophan at position 574 (Yu *et al.*, 2010). So far, three mutations have been described in *A. spica-venti* (Thr, Ala and Ser) at Pro197 and one mutation (Leu) at Trp574 (Balgheim *et al.*, 2007; Hamouzová *et al.*, 2011) and His substitution at Arg377 (Massa *et al.*, 2011).

Mutations occurring at these sites confer different resistance patterns. Substitutions at Ala122 and Ala205 mostly result in resistance to imidazolinone herbicides. The mutation at P197 as well as a different amino acid change has been reported in sulfonylurea-resistant biotypes, Asp376 is responsible for resistance to all classes of ALS inhibitors. It was first shown that the mutation in position 574 confers resistance to both sulfonylureas and imidazolinones using the mutated gene generated by site-directed mutagenesis (Hand *et al.*, 1992). On the other hand, a mutation of Ser653 was first found in imidazolinones-resistant *Arabidopsis* (Haughn and Somerville, 1990), with evolving resistance to pyrimidinylthiobenzoates as well and Gly654 are associated with resistance to imidazolinones solely (Tranel *et al.*, 2010).

It was found that higher plants have a variable number of ALS genes varies from one in *Arabidopsis* (Mazur *et al.*, 1987) to five in *Brassica napus* (Rutledge *et al.*, 1991; Oullet *et al.*, 1992), mainly depending on the level of ploidy: diploid *Arabidopsis thaliana* has one constitutive copy; allotetraploid *Nicotiana tabacum* has two unlinked ALS loci (Keeler *et al.*, 1993); *Zea mays* has two very similar ALS isozymes (Fang *et al.*, 1992); *Brassica napus*, an amphidiploid of *B. campestris* and *B. oleracea*, has five ALS genes. No introns have yet been found in ALS genes.

To determine the molecular basis of resistance, a highly conserved region of the ALS gene containing potential mutation sites was amplified, sequenced and compared between the R and S biotypes. For better targeting of resistant plants for DNA analysis, shoot material of survivors from the R biotypes, following 500 g ha^{-1} sulfomethuron-methyl treatment, were used for DNA extraction. In *A. spica-venti*, it was conferred substitution of alanine in the R biotype for proline in the susceptible biotype at amino acid position 197 (Figure 3). Mutation was found in 10% of tested biotypes. The existence of resistant plants without a mutation at the domains A, B, and D suggests that other mutation(s) can be involved in resistance to sulfonylureas. In some biotypes, mechanism of resistance is probably enhanced metabolism. The mechanism of resistance in the tested populations is not fully clear at present. The complete sequence of the *als* gene of a resistant biotype was gained (GenBank accession number JN646110).

Domain A

	191												203
D36-S	Ala	Ile	Thr	Gly	Glu	Val	**Pro**	Arg	Arg	Met	Ile	Gly	Thr
D36-S	GCC	ATC	ACG	GGG	CAG	GTT	**CCC**	CGC	CGC	ATG	ATC	GGC	ACG
D75-R	–	–	–	–	-C	–	**GCC**	-T	–	–	-A	–	-C
D75-R	–	–	–	–	–	–	**Ala**	–	–	–	–	–	–

Domain B

	593			596
D36-S	Glu	Trp	Glu	Asp
D36-S	CAG	**TGG**	GAG	GAC
D75-R	–	–	–	–

Domain D

	205					210
D36-S	Ala	Phe	Gln	Glu	Thr	Pro
D36-S	GCC	TTC	CAA	GAG	ACG	CCC
D75-R	–	–	-G	–	–	-G

Fig. 3. Nucleotide sequence and inferred amino acid sequences of ALS gene Domain A, B and D from chlorsulforon-resistant (R) and -susceptible (S) *A. spica-venti* biotypes. The codon for amino acid 197 is indicated in bold type. – indicates identical codon to the consensus sequence.

9. Cross resistance

The cross resistance is usually defined as the mechanism (target-site based or non-target-site based) that endows the ability to withstand herbicides from the same of different chemical classes with similar mode of action. We are talking about multiple resistance when a population is resistant to more than one herbicides from different chemical classes (Hall *et al.*, 1994).

As mentioned above, not all ALS-resistant weed biotypes are cross-resistant to all classes of ALS-inhibiting herbicides (Tranel and Wright, 2002) and variable patterns of cross-resistance between ALS-inhibitor classes occur depending on the ALS amino acid position affected and the specific substitution (Shaner, 1999). Some populations of blackgrass and wild oat found in the United Kingdom showed multiple resistance to chlorotoluron, chlorsulfuron, diclofop-methyl, fluazifop-butyl, imazamethabenz-methyl, pendimethalin, simazine, tribenuron and triallate herbicides (Moss, 1990; De Prado and Franco, 2004).

Varying, but very high levels of resistance to chlorsulfuron, sulfosulfuron and iodosulfuron were found, and cross-resistance to all three herbicides was confirmed in *A. spica-venti* from the Czech Republic (Figure 2, Table 5). The resistance to other sulfonylureas was confirmed in the same biotypes. The cross-resistance pattern was obtained by sulfonylurea and triazolopyrimidine herbicides (pyroxsulam), but not by imazethapyr-NH$_4$ while imidazolinones controlled *A. spica-venti* completely. The only population resistant to imidazolinones did not show resistance to other sulfonylureas. All the biotypes tested showed high susceptibility to the ACCase herbicides and PS 2 inhibitors. These results suggested the potential use of the ACCase inhibiting herbicides, as well as isoproturon, in control of ALS resistant biotypes of *A. spica-venti*.

Biotype	Chlorsulfuron 15 g a.i. ha⁻¹	Iodosulfuron 10 g a.i. ha⁻¹	Sulfosulfuron 20 g a.i. ha⁻¹	Mesosulfuron + iodosulfuron 18 g a.i. ha⁻¹	Pyroxsulam 10 g a.i. ha⁻¹	Pinoxaden 30 g a.i. ha⁻¹	Fenoxaprop-P-ethyl 69 g a.i. ha⁻¹	Isoproturon 1000 g a.i. ha⁻¹
				Active ingredient and dose				
1	85	97	90	98	97	100	99	100
2	70	98	99	98	98	99	98	100
3	30	50	55	70	60	100	100	100
4	10	100	91	100	95	100	100	100
5	90	100	100	100	100	100	100	100
6	0	40	20	80	25	100	99	100
7	0	40	25	60	20	100	98	100
8	40	60	40	92	20	100	100	100
9	10	90	5	99	40	100	100	100
10	0	100	98	100	60	100	100	100
S standard	90	95	91	99	98	100	98	100
R standard	40	25	40	70	40	100	94	100

Table 5. Multiple-resistance pattern of *A. spica-venti* tested in whole plant experiment expressed as efficacy (% biomass reduction vs. untreated) 30 days after the treatment.

10. Relative fitness of the resistant and susceptible biotypes

The fitness is often described as an ability of biotype to survive and reproduce in an environment that may or may not include herbicide treatment (Nandula, 2010). Several studies have been taken to assess the differences in resistant and susceptible biotypes to ALS inhibitor with no clear results. Some of them identify significant differences in biomass production, seed production and competitiveness (Christoffoleti *et al.*, 1997), others did not find such differences (Eberlein *et al.*, 1999). The interpretation of the plant fitness comparison is very difficult and the differences may be caused by genetic polymorphism of the biotype rather than by resistance mutation.

Vila-Aiub *et al.* (2009) reviewed that certain gene mutations endowing target site-based herbicide resistance have adverse pleiotropic effects on plant growth and fitness. It is likely that some mutations would impair ALS functionality, and was proven that resistant mutations resulted in higher extracted ALS activity, slightly increased enzyme kinetics and markedly increased sensitivity to feedback inhibition by branched chain amino acids (Yu *et al.*, 2010).

Park *et al.* (2004) observed no differences in competitive ability between resistant and susceptible biotypes of *Bromus tectorum* and it appears that ALS-resistance trait is not associated with growth reductions.

Soukup *et al.* (2006) observed that *Apera* seeds germinated well after maturation, but during after-ripening the germination increased linearly to mid-October (60 %). No differences in germination were found between the resistant and sensitive populations during after-ripening, later the resistant biotype showed slightly lower germination than the susceptible one (Figure 4). Primary dormancy was short and the seeds germinated soon after harvest. Buried seeds are persistent in the soil for only a short time (1 - 4 years) *A. spica-venti* emerges mainly in autumn, best from a depth of a few millimetres, especially under conditions of warm and rainy weather.

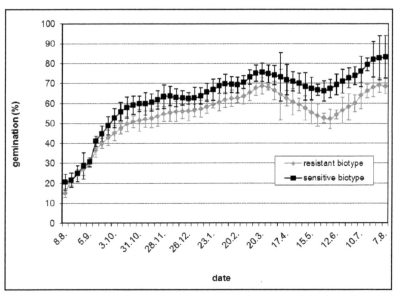

Fig. 4. Germination and loss of primary dormancy of the ALS-resistant and -susceptible biotype of *A. spica-venti* seeds under controlled conditions (20°C, 12 h light) during one year (adopted from the Ph.D. thesis of Náměstek, 2008).

11. Localities with occurrence of resistant biotypes

A. spica-venti occurs frequently in the west and central part of the Czech Republic. Soukup *et al.* (2006) referred that *A. spica-venti* occurs regularly in 80 % of all observed localities and increasing harmfulness was declared by more than 30 % of the inquired farmers and problems with lower efficacy was noticed on 20 % of farms (150 respondents in total). There is no clear pattern of resistance in *Apera* throughout the country (Figure 5); the rapid emergence of the resistant *A. spica-venti* populations was attributed to the lack of herbicide rotation.

Fig. 5. Map of the areas infested with chlorsulfuron-resistant *A. spica-venti* biotypes (in red), data are not derived from the random sampling but based on poor weed control observed by farmers.

12. Forecast of occurrence and level of resistance

The continuous spread of *A. spica-venti* sulfonylurea resistance has been observed since 2003, it can be also attributed to the increasing interest in this grass weed. As seen from Figure 5, the problem is widespread throughout the whole country, but the question how rapidly resistance is spreading was not satisfactorily discussed up to now. During the first years, herbicide resistant plants are present at small numbers and can be incorrectly recognized as escapes. There are suggestions that 30% of a population must be resistant before it is recognized at the field level (Nandula, 2010). At this time, the soil seed bank is established and the control against the resistant weed must be managed integrally. Nevertheless, it may be slowed by using alternative herbicides with more effective modes of action.

Detailed studies on the biochemical basis of resistance in populations of *A. spica-venti* were not undertaken. The recorded actual levels of resistance differed significantly from relatively low (1-fold) and extremely high levels (>200-fold) in terms of resistance factors. These showed that enhanced metabolism existed in some populations but ALS target site resistance occurred in others. From the series of studies, enhanced metabolism could be concluded as the major mechanism of resistance, but target site resistance or mixed resistance (exhibiting both mechanisms) is also present (Hamouzová *et al.*, 2011).

13. Means of control of resistant biotypes

Modern herbicides must provide effective control of *A. spica-venti*, as this is one of the most competitive weeds present in winter cereals in Central and North-west Europe and, if not appropriately controlled, causes serious yield losses in cereal crops. Unfortunately, one of the newest and favourite herbicide groups, the sulfonylureas, is losing its effect due to frequent use. The evolution of resistance against sulfonylureas cannot be underestimated. Their market share is still increasing, and eight new active ingredients have been introduced in the last seven years. Sulfonylureas (chlorsulfuron, iodosulfuron, sulfosulfuron), and ureas (isoproturon, chlortoluron) which create together 80 % of the total usage of herbicides are the most and long-term used herbicides against *A. spica-venti*. On many farms with product failures, there is a tendency to replace sulfonylureas with other herbicides. Early post-emergence herbicides include diflufenican, flufenacet, isoproturon, pendimethalin and prosulfocarb. For later post-emergence application, preferred in integrated weed management, there are only a few products available with different modes of action. ACCase inhibiting herbicides at the moment provide an alternative for resistant biotypes control in winter wheat, but these must be combined with active ingredients against dicotyledonous weeds. The PS2 inhibiting herbicides seem to be effective as well. It is predicted that isoproturon-based products will be restricted soon. It is unlikely that any herbicides based on completely new modes of action will become available for grass weed control within the next few years. The reliance on one herbicide and a shift to another one cannot fully solve the resistance problem.

Crop rotation is considered an effective strategy to manage weed species, in long-term management of particular field, can create unfavourable conditions for a specific weed species and sustain the weed control.

The development of strategies to prevent and manage herbicide resistance should be approached by integrating knowledge from population and evolutionary biology into weed science. Reduction of selection pressure is the main goal of resistance management strategies that rely on herbicide rotations, mixtures and sequences (Neve, 2007).

14. Acknowledgements

This research was funded by projects MSMT6046070901 from the Ministry of Education, Youth and Sports of the Czech Republic and MZE0002700604 from the Ministry of Agriculture of the Czech Republic.

15. References

Anonymous (2010): The agronomic benefits of glyphosate in Europe - Benefits of glyphosate per market use - Review. Monsanto International Sàrl and Monsanto Europe sa, p. 64.

Altschul S.E., Madden T.L., Schäfer A.A., Zhang J., Zhang Z., Miller W., Lipman D.J. (1997): Gapped BLAST and PSI-BLAST: A new generation of protein database search program. Nucl. Acids Res., 25: 3389-3402.

Balgheim N., Wagner J., Gerhards R. (2007): ALS-inhibitor resistant *Apera spica-venti* (L.) Beauv. due to target-site mutation. In: Proc. 14th European Weed Research Society Symposium, Hamar, Norway, 147.

Beckie H.J., Heap I.M., Smeda R.J., Hall L.M. (2000). Screening for herbicide resistance in weeds. Weed Technol., 14: 428-445.

Bradford M. (1976): A rapid and sensitive method for the quantitation of microgram quantities of protein utilizing the principle of protein-dye binding. Analytical Biochemistry, 72: 248-254.

Chodová D., Mikulka J., Kočová M., Janáček J. (1994): Different growth and photosynthetic efficiency of triazine resistant and susceptible biotypes of Poa annua L. Biologia Plantarum, 36, Supplement 1: 341.

Chodová D., Mikulka J., Kočová M., Salava J. (2004): Origin, mechanism and molecular basis of weed resistance to herbicides. Plant Protect. Sci., 40: 151-168.

Chodová D., Salava J. (2004): The evolution and present state of weed resistance to herbicides in the Czech Republic. Herbologia, 5: 11-21.

Chodová D., Salava J., Martincová J., Cvikrová M. (2009): Horseweed with reduced susceptibility to glyphosate found in the Czech Republic. J.Agric.Food Chem., 57: 6957-6961.

Christoffoleti P.J., Westra P., Moore F. III. (1997): Growth analysis of sulfonylurea-resistant and -susceptible kochia (Kochia scoparia). Weed Sci., 45: 691-695.

Délye C., Boucansaud K. (2008): A molecular assay for the proactive detection of target site-based resistance to herbicides inhibiting acetolactate synthase in Alopecurus myosuroides. Weed Res., 48: 97-101.

De Prado R., Franco A. R. (2004): Cross-resistance and herbicide metabolism in grass weeds in Europe: biochemical and physiological aspects. Weed Sci., 52: 441-447.

De Prado R., López-Martínez N., Gonzales-Gutierrez J. (1999): Identification of two mechanisms of atrazine resistance in Setaria faberi and Setaria viridis biotypes. Pesticide Biochem., 67: 114-124.

Duggleby R.G., Pang S.S. (2000): Acetohydroxyacid synthase. Journal of Biochemistry and Molecular Biology, 46: 309-324.

Devine M.D., Shukla A. (2000): Altered target sites as a mechanism of herbicide resistance. Crop Protect. 19: 881-889.

Eberlein C.V., Guttieri M.J., Berger P.H., Fellman J.K., Mallory-Smith C.A., Thill D.C., Baerg R.J., Belknap W.R. (1999): Physiological consequences of mutation for ALS-inhibitor resistance. Weed Sci., 47: 383-392.

Everman W.J., Cassandra R.M, Burton J.D., York A.C., Wilcut J.W. (2009): Absorption, translocation and metabolism of 14C-glufosinate in glufosinate-resistant corn, goosegrass (Eleusine indica), large crabgrass (Digitaria sanquinalis) and slickedpod (Senna obtusifolia). Weed Sci., 57: 1-5.

Fang L.Y., Gross P.R., Chen P.H., Lillis M. (1992): Sequence of two acetohydroxyacid synthase genes from Zea mays. Plant Mol. Biol., 18: 1185–1187.

Frey J.E., Müller-Schärer H., Frey B., Frey D. (1999): Complex relation between triazine susceptible phenotype and genotype in the weed Senecio vulgaris may be cause by chloroplast DNA polymorphism. Theor. Appl. Genet., 99: 578-586.

Gadamski G., Ciarka D., Gawronski S.W. (1996): Molecular survey of Polish resistant biotypes of weeds. In: Proc. 2nd Int. Weed Control Congress, Copenhagen, pp. 547-550.

Gasteiger E., Gattiker A., Hoogland C., Ivanyi I., Appel R.D., Airoch A.B. (2003): ExPASY: the proteomics server for in-depth protein knowledge and analysis. Nucl. Acids Res., 31: 3784-3788.

Gawronski S.W., Sugita M., Sugiura M. (1992): Mutation of *psb*A gene in herbicide resistant populations of *Erigeron canadensis*. In: Murata N. (Ed.): Research in Photosynthesis, pp. 405-407. Kluwer Academic Publishers, Dordrecht.

Hall L.M., Holtum J.A.M., Powles S.B: (1994): Mechanisms responsible for cross resistance and multiple resistance. In: Holtum J.A.M. and Powles S.B., eds. Herbicide Resistance in Plants: Biology and Biochemistry. Boca Raton, FL: Lewis Publishers, pp. 243-261.

Hamouzová K., Soukup J., Jursík M., Hamouz P., Venclová V., Tůmová P. (2011): Cross-resistance to three frequently used sulfonylurea herbicides in populations of *Apera spica-venti* from the Czech Republic. Weed Res., 51: 113-122.

Hand J.M., Singh B.K., Chaleff R.S. (1992): American Cyanamid Company: Eur. Pat. Appl. EP 492113.

Heap I. (2000): International survey of herbicide resistant weeds. Internet. 10 February 2000. Available at www.weedscince.com.

Heap I. (2011): International survey of herbicide resistant weeds. Internet. 16 June, 2011. Available at www.weed-science.com.

Hirschberg J., McIntosh L. (1983): Molecular basis for herbicide resistance in *Amaranthus hybridus*. Weed Sci., 43: 175-178.

Holá D., Kočová M., Körnerová M., Sofrová D., Sopko B. (1999): Genetically based differences in photochemical activities of isolated maize (*Zea mays* L.) mesophyll chloroplasts. Photosynthetica, 36:187-268.

Haughn G. W., Somerville C. R. (1990): A mutation causing imidazolinone-resistance maps to the Csr1 locus of *Arabidopsis thaliana*. Plant Physiol., 92: 1081–1085.

Kaundun S.S., Windass, J.D. (2006) Derived cleaved amplified polymorphic sequence, a simple method to detect a key point mutation conferring acetyl CoA carboxylase inhibitor herbicide resistance in grass weeds. Weed Res., 46: 34-39.

Kaundun S.S., Hutchings S.-J., Dale R.P., Bailly G.C., Glanfield P. (2011). Syngenta RISQ test: a novel in-season method for detecting resistance to post-emergence ACCase and ALS inhibitor herbicides in grass weeds. Weed Res., 51: 284-293.

Keeler S.J., Sanders P., Smith J.K., Mazur B.J. (1993): Regulation of tobacco acetolactate synthase gene expression. Plant Physiol., 102: 1009–1018.

Körnerová M., Holá D., Chodová D. (1998): The effect of irradiance on Hill reaction activity of atrazine- resistant and -susceptible biotypes of weeds. Photosynthetica 35, 265-268.

Mallory-Smith C.A., Hendrickson P., Mueller-Warrant G.W. (1990): Identification of herbicide resistant prickly lettuce (*Lactuca serriola*). Weed Technol., 4: 163-168.

Massa D., Krenz B., Gerhards R. (2011): Target-site resistance to ALS-inhibiting herbicides in *Apera spica-venti* populations is conferred by documented and previously unknown mutations Weed Res., 51: 294-303.

Mazur B.J., Chui C.F., Smith J.K. (1987): Isolation and characterization of plant genes coding for acetolactate synthase, the target enzyme for two classes of herbicides. Plant Physiol., 85: 1110-1117.

McCourt J.A., Duggleby R.G. (2006): Acetohydroxyacid synthase and its role in the biosynthetic pathway for branched-chain amino acids. Amino Acids, 31: 173-210.

Monteiro I., Rocha F. (1992): Study of a survey of weed biotypes resistant to atrazine. In: Proc. 1992 Congress Spanish Weed Science Society, 315-319.

Moss S.R. (2002): Herbicide-resistant weeds. In: Nylor R.E.L., editor. Weed management handbook. 9th edition. Oxford (UK): Blackwell Science Ltd for British Crop Protection Council. pp. 225-252.

Moss S.R. (1990): Herbicide cross-resistance in slender foxtail (*Alopecurus myosuroides*). Weed Sci., 38: 492-496.

Moss S.R., Perryman S.A.M., Tatnell L.V. (2007): Managing herbicide-resistant blackgrass (*Alopecurus myosuroides*): Theory and Practice. Weed Technol., 21:300-309.

Náměstek J. (2008): Biologie a geografické rozšíření chundelky metlice (*Apera spica-venti* ((L.) P.B.) v ČR a studium její senzitivity k vybraným herbicidům. Ph.D. thesis, Czech University of Life Sciences, Prague, Czech Republic, 111 pp.

Nandula V.K. (eds) (2010): Glyphosate resistance in crops and weeds: history, development, and management, John Wiley & Sons, Inc., Hoboken, New Jersey. pp. 321.

Neve P. (2007): Challenges for herbicide resistance evolution and management: 50 years after Harper. Weed Res., 47: 365–369.

Nielsen O.K., Ritz C., Streibig J.C. (2004): Non-linear mixed-model regression to analyze herbicide dose-response relationships. Weed Technol., 18: 30–37.

Nováková K., Salava J., Chodová D. (2005): Biological characteristics of an atrazine resistant common groundsel (*Senecio vulgaris* L.) biotype and molecular basis of the resistance. Herbologia, 6: 65-74.

Nováková K., Soukup J., Wagner J., Hamouz P., Náměstek J. (2006): Chlorsulfuron resistance in silky bent-grass (*Apera spica-venti* (L.) Beauv.) in the Czech Republic. J. Pl. Dis. Protect., Special issue XX: 139-146.

Ort D.R., Ahrens W.H., Martin B., Stoller, E. (1983): Comparison of photosynthetic performance in triazine-resistant and -susceptible biotypes of *Amaranthus hybridus*. Plant Physiol. 72: 925-930.

Oullet T., Rutledge R.G., Miki B.L. (1992): Members of the acetohydroxyacid synthase multigene family of *Brassica napus* have divergent patterns of expression. Plant J., 2: 321-330.

Park K.W., Mallory-Smith C.A., Ball D.A., Mueller-Warrant G.W. (2004): Ecological fitness of acetolactate synthase inhibitor-resistant and -susceptible downy brome (*Bromus tectorum*) biotypes. Weed Sci., 52: 768-773.

Powles S.B., Yu Q. (2010): Evolution in action: plants resistant to herbicides. Annual Review of Plant Biology, 61: 317-347.

Preston C., Roush R.T., Powles S.B. (1999): Herbicide resistance in weeds of southern Australia: why are we the worst in the world? In: Bishop A., Boersma C., Barnes C.D. (eds). 12th Australian Weeds Conference: Papers and Proceedings, Tasmanian Weed Society: Davenport, Australia, pp 454–459.

Preston C., Powles S. B. (2002): Evolution of herbicide resistance in weeds: initial frequency of target site-based resistance to acetolactate synthase-inhibiting herbicides in *Lolium rigidum*. Heredity: 88, 8-13.

Ray T.B. (1984): Site of action of chlorsulfuron. Plant Physiol., 75: 827 - 831.

R Development Core Team (2008): R: A Language and Environment for Statistical Computing. R Foundation for Statistical Computing, Vienna, Austria. ISBN 3-900051-07-0. Available at: http://www.R-project.org.

Ritz C., Cedergreen N., Jensen J.E., Streibig J.C. (2006): Relative potency in non-similar dose-response curves. Weed Sci., 54: 407–412.

Robinson D., Greene D. (1976): Metabolism and differential susceptibility of crabgrass and witchgrass to simazine and atrazine. Weed Sci., 24:500-504.

Rutledge R., Ouellet T., Hattori J., Mlki, B.L. (1991): Molecular characterization and genetic origin of the *Brassica napus* acetohydroxyacid synthase multigene family. MOl. Gen. Genet., 229: 31-40.

Salava J., Chodová D., Mikulka J. (2004): Molecular basis of acetolactate synthase-inhibitor resistance in Czech biotypes of kochia. J. Pl. Dis. Protect., Special issue XIX: 915-919.

Salava J., Kočová M., Holá D., Rothová O., Nováková K., Chodová D. (2008): Identification of the mechanism of atrazine resistance in a Czech large crabgrass biotype. J. Pl. Dis. Protect., Special issue XXI: 101-104.

Shaner D. L. (1999): Resistance to acetolactate synthase (ALS) inhibitors in the United States: history, occurrence, detection and management. Weed Sci., 44: 405–411.

Shimizu T., Nakayama I., Nagayama K., Miyazawa T., Nezu Y. (2002): Herbicide Classes in Development: Mode of Action-Targets-Genetic Engineering-Chemistry, ed. by Böger P., Wakabayashi K. and Hirai K., Springer-Verlag, Berlin Heidelberg, pp. 1-41.

Slavíková L., Mikulka J., Kundu J. K. (2011): Tolerance of blackgrass (*Alopecurus myosuroides*) to sulfonylurea herbicides in the Czech Republic. Plant Protect. Sci., 47: 55-61.

Soukup J., Nováková K., Hamouz P., Náměstek J. (2006): Ecology of silky bent grass (*Apera spica-venti* (L.) Beauv.), its importance and control in the Czech Republic J. Pl. Dis. Protect., Special Issue XX: 73-80.

Sundby C., Chow W.S., Anderson J.M. (1993): Effects on photosystem II function, photoinhibition, and plant performance of the spontaneous mutation of serin264 in photosystem II reaction center D1 protein in triazine resistant *Brassica napus* L. Plant Physiol. 103: 105-113.

Tranel P.J., Wright T.R., Heap I. (2010): ALS mutations from herbicide-resistant weeds. Available at: http://www.weedscience.com (last accessed 21 June 2010).

Tranel P.J., Wright T.R. (2002): Resistance of weeds to ALS-inhibiting herbicides: what have we learned? Weed Sci., 50: 700–712.

Vila Aiub M., Neve P.B., Powles S.B. (2009): Fitness costs associated with evolved herbicide resistance alleles in plants. New Phytologist, 184: 751-767.

Wagner J., Laber B., Menne H., Kraeruthmer H. (2008): Rapid analysis of target-site resistance in blackgrass using Pyrosequencing® technology. The 5th International Weed Science Congress, 23 - 27 June, Vancouver, Canada.

Yu Q., Han H., Vila-Aiub M.M., Powles S.B. (2010): AHAS herbicide resistance endowing mutations: effect on AHA functionality and plant growth. J. Exp. Bot., 61 (14): 3925-3934.

6

The Use of Herbicides in Biotech Oilseed Rape Cultivation and in Generation of Transgenic Homozygous Plants of Winter Oilseed Rape (*Brassica napus* L.)

Teresa Cegielska-Taras[1] and Tomasz Pniewski[2]
[1]Plant Breeding and Acclimatization Institute,
National Research Institute, Division Poznań,
[2]Institute of Plant Genetics Polish Academy of Science, Poznań,
Poland

1. Introduction

Oilseed rape (rapeseed, *Brassica napus* L.) is one of the most important crop plants, taking the third place after palm oil and soybean among oil plants and the fifth among economically important crops, following rice, wheat, maize and cotton (Snowdon et al., 2007). Oilseed rape is cultivated primarily in temperate regions, where climatic conditions prevent growth of most other oilseed crops. Spring cultivars of oilseed rape dominate in the USA, Canada, Australia and China, while winter varieties are cultivated mainly in northern European countries (Great Britain, France, Germany, Poland, Scandinavia, etc.).

Over the past few decades, a number of crucial results in oilseed rape breeding research have been found. Improvement of oilseed rape by reducing erucic acid content in oil and glucosinolates in meal, were the two most prominent milestones, which contributed to the obtaining of '00' cultivars (also known as a canola), and subsequent world-wide expansion in the production of oilseed rape (Vollmann & Rajcan, 2009). Therefore, rapeseed oil has become one of the basic cooking oils and provides ca. 10 - 15% of total world production of vegetable oils. Moreover, rapeseed oil, free of anti-nutritional substances, has a higher dietary value than many other oils due to the lowest amount of saturated fats among vegetable oils, high oleic acid content and a favourable ratio of unsaturated fatty acids (linoleic and linolenic acids), as well as sterols and fat-soluble vitamins. Rapeseed oil is also used for varnish production and is increasingly being exploited as a bio-fuel. Meal remained after oil extraction as well as specially grinded seeds are valuable high-protein fodder.

Therefore, oilseed rape is an object of extensive efforts of breeding programmes to generate high yielding cultivars that are resistant to several biotic and abiotic stress conditions. Although breeding programmes based on natural genetic variation in *Brassica napus* are crucial, biotechnological approaches are essential as a complement to traditional procedures. Several methods, based on tissue cultures, e.g. interspecific and intergeneric hybridization combined with embryo rescue techniques, protoplast fusion or production of doubled haploids (DH) have been successfully exploited in oilseed rape breeding programmes for over 30 years. Since the last two decades, genetic transformation has proven to be a significant method employed in commercial oilseed rape improvement.

2. Gene transfer in improvement of oilseed rape

Due to its economical importance, *Brassica napus* L. was among the first genetically transformed crops in the history of genetically transformed plants (Maheshwari et al., 2011). Various physical techniques of gene transfer to oilseed rape have been tested, including microprojectile bombardment (Chen & Beversdorf, 1994; Fukuoka et al., 1998), electroporation (Chapel & Glimelius, 1990), microinjection (Jones-Velleneunes et al., 1995) and PEG-mediated DNA uptake (Golz et al., 1990). However, *Agrobacterium tumefaciens*-mediated method is the preferred choice since it is an easy and effective way to transfer foreign genes to plant cell. (Cardoza and Stewart, 2003; Friedt & Snowdon, 2009; Jonoubi et al., 2005; Moghaieb et al., 2006; Maheshwari et al., 2011; Sharma et al., 2005; Takahata et al., 2005). Many factors have been investigated to elaborate transformation protocols, including oilseed rape varieties, *Agrobacterium tumefaciens* and *Agrobacterium rhizogenes* strains, binary vectors, type of explant (leaf pieces, cotyledons, hypocotyls, stem segments etc.) and the selection system such as antibiotics, herbicides etc. (De Block et al., 1989; Pechan, 1989; Poulsen, 1996; Wang J. et al., 2005; Wang W.C. et al., 2003; Wang Y.P. et al., 2005). Beside efficient transformation methods, crucial parameters for the use of transgenic crops are stable inheritance of the transgene and its expression and harmonization of input traits with the host genome and physiology. Transgenic cultivars or production lines with proper characteristics have to be selected from among hundreds and thousands primary transformants.

Nowadays, with the development of transformation techniques, targeting precisely defined genes directly into the biological characteristics of oilseed rape, biotechnology has become very important for the improvement of oilseed rape breeding. Although, so far only spring oilseed rape was successfully transformed, up till now 40 genes (transgenes) have been introduced into the oilseed rape genome (ISAAA, 2011). Transformation has been used for basic research purposes, for example deciphering the functioning of the myrosinase system (Troczyńska et al., 2003). Yet, most of reports on transgenic, i.e. genetically modified (GM) oilseed rape, were referred to new utilitarian and agricultural traits such as resistance to herbicides, viruses, fungi, insects and environmental factors such as drought, salinity, as well as male sterility and changes in the composition of fatty acids and proteins in the seeds (Dill, 2005; Kahrizi et al., 2007; McGloughlin, 2010; Poulsen, 1996; Senior & Bavage, 2003; Sharma et al., 2005; Wang J. et al., 2005; Zhong, 2001). However, the predominant genetic modification of commercially cultivated transgenic oilseed rape, also called GM canola, is herbicide resistance (Table 1).

3. Herbicides in GM canola agriculture

The first commercial launch of GM plants tolerant to herbicide, which was glyphosate-resistant soybean in 1996, signalled the beginning of a new era in weed management in row crops. Since that time several crops, mainly soybean, maize, cotton and canola, have been transformed that have allowed crop applications of many classes of herbicide chemistries. Such crops tolerate herbicides which completely eliminate weeds. As a result, one-two sprays (e.g. after crop emergences) per vegetative season reduce the number of agronomical interventions, labour cost, fuel, machine consumption etc., and consequently generate lower economical costs. However, there is still ongoing discussion on the risk of uncontrolled release of herbicide-resistant crops into the environment and a potential danger of spreading of genes determining resistance to herbicides which could lead to the rise of herbicide-tolerant weeds. Nevertheless protests of some organisations such as Greenpeace etc., generally low public acceptance and governmental stands in some countries (mostly European ones), it appears that GM crops will continue to grow in species and cultivar number and in hectares planted, ca. 5-10% per year. In 2010, GM crops were cultivated on an area of 148 mln ha, in 29 countries. Dominant producers of GMO (Genetically Modified Organism) are the USA, Argentina Brazil, Canada and China, but the number of countries admitting GMO is growing, mainly in Asia, Africa and Latin America, and even in European Union, GM plants are cultivated in 9 countries on over 100,000 ha. So far, field trials of GMO have been conducted for almost 140 species in 11,400 experiments and approved commercial events have reached 80 cultivars among 16 species. The largest part of entire GMO area is occupied by soybean, maize, cotton and canola. Since the beginning of GMO agriculture, herbicide resistance is the predominant genetic modification. Nowadays, more than 70% of cultivated GM plants or 61% of GM occupied area over the world are herbicide resistant (ISAAA, 2011).

The herbicides utilized in the cultivation of GM canola and other crops are characterized by non-selective and systemic mode of action and foliar penetration. There are three main groups of these herbicides, distinguished by different active constituents: glyphosate (N-(phosphonomethyl)glycine) (Fig. 1a), oxynils (Fig. 1b) and phosphinothricin (glufosinate, glufosinate ammonium) (Fig. 1c).

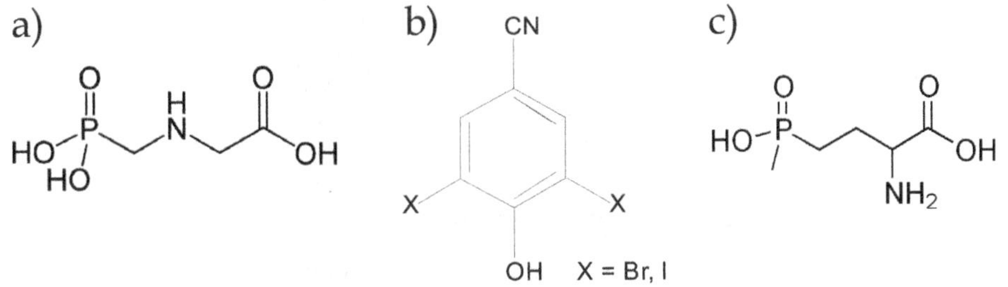

Fig. 1. Chemical structure of active constituents of herbicides used in GM canola agriculture.
a) glyphosate - N-(phosphonomethyl)glycine; IUPAC name -
2-[(phosphonomethyl)amino]acetic acid; b) oxynil; IUPAC name - 4-hydroxy-3,5-dihalogenobenzonitrile, halogene – bromine or iodine; c) phosphinothricin – glufosinate, glufosinate ammonium; IUPAC name - 2-amino-4- (hydroxy(methyl)phosphonoyl)butanoic acid.

Glyphosate inhibits the enzyme 5-enolpyruvylshikimate-3-phosphate synthase (EPSPS), which catalyzes the reaction of 5-enolpyruvyl-shikimate-3-phosphate (ESP) formation. ESP is a precursor for the aromatic amino acids: phenylalanine, tyrosine and tryptophan, used as protein building blocks and in synthesis of such essential metabolites as folic acid and quinones. Glyphosate-based herbicides such as Roundup®, are produced by Monsanto. Roundup Ready™ crops, i.e. transgenic plants resistant to glyphosate, contain genes originated from *Agrobacterium tumefaciens* strain C4 and encoding the C4 EPSPS form insensitive to glyphosate, and/or the *gox* gene encoding glyphosate oxidase derived from *Ochrobactrum anthropi*.

Oxynils are lethal to plants, as they inhibit photosynthesis by blocking electron transport at photosystem II. Herbicides containing oxynils - Actril®, Totril®, are provided by Bayer Crop Science Company. Resistant crops contain *bxn* gene from *Klebsiella ozaenae*, which encodes a nitrilase enzyme that hydrolyzes oxynils to non-toxic compounds.

Phosphinothricin or glufosinate (glufosinate ammonium) is an active ingredient in several herbicides -- Basta®, Rely®, Finale®, Ignite®, Challenge® and Liberty® and other proprietary pesticides, produced by Bayer Crop Science Company. Glufosinate is a glutamine synthetase inhibitor that binds to the glutamate site of the enzyme. Consequently, ammonia builds up to toxic concentration, next pH increases to alkaline level as well as a cessation of photosynthesis resulting in the lack of glutamine and plant death. Resistance of transgenic plants to glufosinate is due to the activity of phosphinothricin N-acetyltransferase (PAT) encoded by the gene *bar* from *Streptomyces hygroscopicus* or *pat* from *Streptomyces viridochromogenes*. This is the most common resistance system used in GM canola production.

Herbicide tolerance is the predominant trait introduced into commercially cultivated transgenic oilseed rape (Table 1). Among 21 events approved, 19 are herbicide-resistant. Beside the above-mentioned economical reasons, herbicide selection system is also exploited due to GMO safety regulations and health requirements, (e.g. Regulation EC No. 1829/2003), which forbid the use of genes determining antibiotic resistance in GM plants utilized as food and feed sources. Herbicide resistance in GM canola was introduced either as a single trait or combined with other agronomical features (Table 1). Nowadays the cultivation area of GM canola constitutes ca. 7.5 mln ha which is 5% of the total GMO global area. The area occupied by GM canola reached 18-20% of the entire oilseed rape area and is distributed mainly in the USA, Canada and China (ISAAA, 2011). However, it should be stressed that so far GM canola events released to agriculture have been obtained only for spring varieties of oilseed rape. Therefore, a transformation method of winter oilseed rape is necessary due to its particular importance in countries of moderate and cool climate.

4. Transgenic homozygous herbicide-resistant winter oilseed rape

Production of transgenic winter oilseed rape, especially herbicide-resistant one, has been delayed in comparison to spring varieties, due to the recalcitrance of winter varieties of *Brassica napus* to transformation via *Agrobacterium* spp. or other techniques. So far transformation attempts have shown very limited success, despite using selection system based on *npt II* gene, conferring resistance to the mildly acting antibiotic kanamycin (Grison et al., 1996; Poulsen, 1996; Takahata et al., 2005; Wang Y.P. et al., 2005). A major constraint for the advance of transgenic breeding has been limitations of initial regeneration systems as well as problems in combining regeneration and transformation within the same cells. The choice of cells and organs as recipients of introduced gene and an efficient system for selection of cells containing an integrated foreign gene in genome, are essential elements of successful plant transgenesis. For the obtained successful gene transfer, cells or tissues have to represent a high regenerative capacity to develop complete and normal transgenic organism. Therefore, it is assumed that under specific conditions, obtaining of transgenic winter oilseed rape is possible.

Event	Description
GT200 (RT200) (MON-89249-2) Roundup Ready™ Canola Developer: Monsanto	Canola tolerant to the herbicide glyphosate produced by inserting genes encoding the enzymes EPSPS and GOX.
RT73 (GT73) (MON-ØØØ73-7) Developer: Monsanto	Glyphosate herbicide tolerant canola (cv. Westar) produced by inserting the *epsps* gene encoding the enzyme EPSPS and GOX.
HCN10 Liberty-Link™ Independence Developer: Aventis Crop Science	HCN10 (Independence) is an open pollinated canola line, which is tolerant to the glufosinate-ammonium (phosphinotricin), the active constituent of the non-selective broad-spectrum herbicides Basta®, Finale®, Buster®, Harvest® and Liberty®.
PGS2 (MS1 x RF2) (B91-4 x B94-2) (ACS-BNØØ4-7 x ACS-BNØØ2-5) Developer: Aventis Crop Science	The canola lines MS1 and RF2 were developed using genetic engineering techniques to provide a pollination control system for the production of hybrid oilseed rape (MS1xRF2) expressing male sterility and tolerance to glufosinate ammonium. The novel hybridization system involves the use of two parental lines, a male sterile line MS1 and a fertility restorer line RF2. The transgenic MS1 plants do not produce viable pollen grains and cannot self-pollinate. In order to completely restore fertility in the hybrid progeny, line MS1 must be pollinated by a modified plant containing a fertility restorer gene, such as line RF2. The resultant F1 hybrid seed, derived from the cross between MS1 x RF2, generates hybrid plants that produce pollen and are completely fertile.
OXY-235 (ACS-BNØ11-5) Navigator™ , Compass™ Canola Developer: Aventis Crop Science	Canola (variety Westar) tolerant to the oxynil herbicides created through insertion of the *bxn* gene encoding a nitrilase enzyme that hydrolyzes oxynils to non-toxic compounds.
MS8 (ACS-BNØØ5-8) InVigor™ Canola Developer: Bayer Crop Science	Canola with male-sterility system displaying glufosinate herbicide tolerance. Contains the barnase gene from *Bacillus amyloliquefaciens* and the *bar* gene encoding PAT. Also contains the *npt II* gene conferring resistance to the antibiotic kanamycin.
PHY14, PHY35, PHY36 Developer: Bayer Crop Science	These lines are high yielding fertile hybrids and tolerant to the herbicide glufosinate-ammonium, used for the selection of the transformants.
RF3 (ACS-BNØØ3-6) InVigor™ Canola Developer: Bayer Crop Science	Canola fertility restoration system displaying glufosinate herbicide tolerance. Contains the *barstar* gene from *Bacillus amyloliquefaciens*, and the *bar* gene encoding PAT to confer tolerance to the

	herbicide glufosinate (phosphinothricin).
T45 (HCN28) (ACS-BNØØ8-2) InVigor™ Canola Developer: Bayer Crop Science	Glufosinate tolerant canola with insertion of the *pat* gene, conferring tolerance to glufosinate ammonium herbicide.
MS8 x RF3 (ACS-BNØØ5-8 x ACS-BNØØ3-6) InVigor™ Canola Developer: Bayer Crop Science, Aventis Crop Science (AgrEvo)	Canola with male-sterility, fertility restoration, pollination control system displaying glufosinate herbicide tolerance.
PGS1 (MS1(B91-4) x RF1(B93-101)) (ACS-BNØØ4-7 x ACS-BNØØ1-4) MS1 x RF1 Developer: Bayer Crop Science, Aventis Crop Science (AgrEvo)	Canola with male-sterility, fertility restoration, pollination control system, and glufosinate herbicide tolerance. MS1 line contained the barnase gene from *Bacillus amyloliquefaciens* (with pTa 29 pollen specific promoter from *Nicotiana tabacum*). RF1 line contained the *barstar* gene from the same bacteria with anther-specific promoter, and both lines contained the *bar* gene encoding PAT (with PSsuAra promoter from *Arabidopsis thaliana*) to confer tolerance to the herbicide glufosinate. Also includes *npt II* gene conferring resistance to the antibiotic kanamycin.
Topas 19/2, HCN92 (ACS-BNØØ7-1, HCN92) Liberty-Link™ Innovator Developer: Bayer Crop Science, Aventis Crop Science (AgrEvo)	Glyphosate herbicide tolerant canola produced by inserting the pat gene conferring tolerance to glufosinate ammonium herbicide and *npt II* conferring resistance to the antibiotic kanamycin.
MPS961, MPS962, MPS963, MPS964, MPS965 **Phytaseed™** Canola Developer: BASF AG	Canola which expresses a phytase gene derived from *Aspergillus niger*. Altered quality product -- canola seeds contain higher content of a usable form of inorganic phosphorus released from phytic acid.
23-18-17, 23-198 (CGN-89111-8, CGN-89465-2) **Laurical™ Canola** Developer: Calgene Inc.	Lines 23-18-17 and 23-198 are high laurate- and myristate-containing canola produced by inserting a thioesterase (te) encoding gene from the California bay laurel (*Umbellularia californica*). The *npt II* gene confers resistance to the antibiotic kanamycin.

Abbreviations:

EPSPS -- 5-enolypyruvylshikimate-3-phosphate synthase non-sensitive to glyphosate, encoded by *epsps* gene from *Agrobacterium tumefaciens* strain C4

GOX – glyphosate oxidase, encoded by the *gox* gene from *Ochrobactrum anthropi*

PAT - phosphinothricin N-acetyltransferase, encoded by the *bar* gene from *Streptomyces hygroscopicus* or the *pat* gene from *Streptomyces viridochromogenes*

npt II – gene encoding neomycin phosphotransferase II

Table 1. GM canola events released to agriculture (on the basis of ISAAA, 2011a; Genescan, 2011).

One of the first successful attempts to obtain transgenic herbicide-resistant winter oilseed rape was the transformation of haploid microspore-derived embryos (MDEs) using hypervirulent *Agrobacterium tumefaciens* EHA105 strain (Cegielska et al., 2008). In this method, two directions of oilseed rape improvement and new cultivar production have been combined – doubled haploids and genetic modification.

Doubled haploids, obtained by tissue cultures of microspores and microspore-derived embryos (MDEs), show uniformity and complete homozygosity, thus they are particularly useful for breeding programmes and have been extensively used to produce homozygous breeding lines and cultivars of oilseed rape (Cegielska et al., 2002; Friedt and Zarhloul, 2005; Takahata et al., 2005). DH plants can also be efficiently used in basic studies, for example in genetic analysis and genetic transformation (Cegielska-Taras et al., 2008; Fukuoka et al., 1998; Huang, 1992; Jardinaud et al., 1993; Kazan et al., 1997; Nehlin et al., 2000; Oelck et al., 1991; Swanson & Erickson, 1989; Takahata et al., 2005; Troczyńska et al., 2003). It is postulated that in the process of genetic transformation, the gene introduced into the haploid genome, in this case MDE and subsequent chromosome doubling (diploidization), it will give rise to homozygous transgenic oilseed rape (Dorman et al., 1998; Huang, 1992).

Herbicide resistance as a marker trait of transgenic winter oilseed rape was chosen due to three reasons. First of all, glufosinate ammonium (phosphinothricin) was able to effectively select transformants of winter oilseed rape in contrast to kanamycin or other antibiotics. Moreover herbicide-tolerant winter oilseed rape conforms GMO regulations and has potential utilitarian value. Appropriate binary vector pKGIB was constructed (Fig. 2), which contained two chimeric genes (transgenes) inside a DNA fragment transferred into plant cells (T-DNA): the *bar* marker gene coding for phosphinothricin N-acetyltransferase under the control of constitutive NOS promoter (nopaline synthase originated from *Agrobacterium tumefaciens* C58 strain) and g7 terminator and the *uid* gene with an intron encoding β-glucuronidase (GUS), under the control of constitutive 35S promoter (35S RNA of Cauliflower Mosaic Virus) and NOS terminator, which was used as a reporter gene.

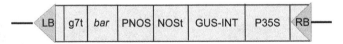

Fig. 2. Organisation of T-DNA of binary plasmid pKGIB. P35S = CaMV 35S promoter; PNOS = nopaline synthase promoter; GUS-INT = gene of β-glucuronidase with an intron; *bar* = gene of phosphinotricin N-acetyltransferase; NOSt = nopaline synthase terminator; g7t = g7 terminator; LB and RB = left and right T-DNA border sequences. (Cegielska et al., 2008)

Hypervirulent *Agrobacterium tumefaciens* EHA105 carrying the vector, after treatment in minimal medium with acetosyringone as an inducer of transformation process, was able to transfer T-DNA into slightly injured and pre-chilled MDEs. Although both oilseed rape microspores and MDEs characterise high regeneration potential (Fig. 3) (Cegielska et al., 2002, Friedt and Zarhloul 2005; Takahata et al., 2005), only MDEs were able to regenerate post *Agrobacterium* inoculation (Fig. 4). Plant regeneration under selection conditions, i.e. in tissue cultures on media supplemented with 10 mg/L of glufosinate ammonium, occurred via somatic embryogenesis, similarly to the usual regeneration method in MDEs during regular DH production (Fig. 5a). Nine putative haploid transformants were regenerated from 100 primary explants. The plants were completely or partially resistant to glufosinate-based herbicide Basta® as observed in leaf-painting test and expressed the reporter gene at satisfactory level (Figs. 5b, 5c).

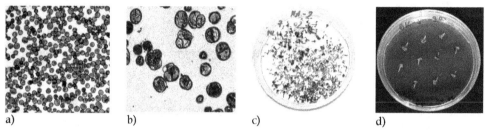

a)   b)   c)   d)

Fig. 3. Development of haploid microspore-derived embryos (MDEs) of winter oilseed rape. a) microspores (pollen grains); b)cellular divisions in regenerating microspores: c) culture of microspore-derived embryos; d) pre-chilled 21day-old microspore-derived embryos at the stage suitable for *Agrobacterium*-mediated transformation.

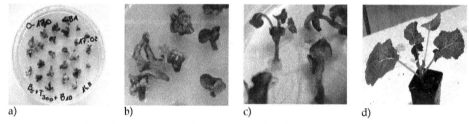

a)   b)   c)   d)

Fig. 4. Regeneration of transgenic haploid winter oilseed rape. a) transformed MDEs on a selection medium supplemented with 10 mg/L of glufosinate ammonium; b) plantlet development from transformed MDEs; c) regenerated and rooted transformants before transferring into soil; d) a transformant adapted to *ex vitro* conditions before colchicine treatment and vernalization.

a)   b)   c)

Fig. 5. Expression of transgenes in transformed MDEs and regenerated plants of winter oilseed rape. a) Localization of expression of the GUS reporter gene in MDE apical bud 10 days after transformation using *A. tumefaciens*. Putative sites of secondary bud formation are indicated by arrows. b) GUS expression (blue staining) in the meristematic region of transformed MDEs (top) and control embryos (bottom). c) Herbicide test – painted leaves (clamped, indicated by arrows) 5 days after herbicide application: control rape DH O-120 (left) and transgenic plants, i.e. partially resistant plant T-25 (centre) and completely resistant T-26 (right). (based on Cegielska et al., 2008)

However, after colchicine treatment and diploidization (Fig. 6a), only one primary transformant (T_0 generation) set vital seeds. In contrast to others characterised by one integration site, this fertile transformant contained two loci of T-DNA integrated in the genome (Fig. 6b). Nevertheless, all its progeny plants (T_1 generation) revealed the presence of *bar* gene present in the genome (Fig. 6c), were diploid and tolerant to herbicide spray.

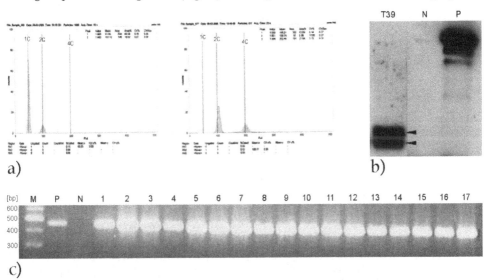

a)   b)

c)

Fig. 6. Molecular analysis of transgenic winter oilseed rape. a) Determination of ploidy level of regenerated transgenic oilseed rape plants by flow cytometry analysis of DNA content (1C, 2C and 4C): haploid T39 transformant (left) and control diploid plant (right).
b) Southern-blot analysis of transgenic plant of DH winter oilseed rape, 25 μg of DNA cut with *Eco*R I. Lanes: T-39 = DH transgenic plant; N = nontransgenic plant, negative control; P = plasmid pKGIB, positive control. c) PCR analysis of the *bar* gene in T_1 plants of DH T-39 transformant of oilseed rape DH line O-120, Lanes: M = DNA size marker; P = plasmid pKGIB, positive control; N = untransformed O-120 plant, negative control; 1-17 = T_1 plants of DH T-39 transformant. (based on Cegielska et al., 2008)

While only one reliable transgenic plant was obtained in the first trial, following experiments verified the *Agrobacterium*-mediated MDE transformation as useful method of gene transfer. For every transformation round, 3-4 fertile transformants were obtained and progeny plants were resistant to Basta® as well as expressed introduced genes (unpublished). Although verification is still required for transgene functionality (including potential gene silencing) in the next generations, so far results strongly indicate that the repeatable method of winter oilseed rape transformation is worked out.

It may be assumed that contrary to routinely produced hemizygotic transformants, transgenic homozygotes will stably express integrated genes due to reduction or even elimination of at least one reason of gene silencing. A two-allelic transgene will probably resemble natural plant genes in more extent than an one-allelic transgene in a hemizygote. This aspect of both theoretical and practical significance is planned to be investigated in the future. However, *Agrobacterium*-mediated MDE transformation makes such research possible.

5. Conclusion

The presented method of winter oilseed rape MDE transformation using *Agrobacterium* appears promising. It allows for obtaining truly genetically modified winter oilseed rape of potential practical application. Main advantage of this method is that, as a result of chromosome duplication in transformed haploid, the introduced trait can be evaluated in a single step in a transgenic homozygote. Production of transgenic homozygous oilseed rape provides unique material for further studies of inheritance and functionality of introduced genes trough subsequent generation, for both basic research, breeding programmes and utilitarian purposes.

6. References

Cardoza, V. & Stewart, C.N. (2003). Increased *Agrobacterium*-mediated transformation and rooting efficiencies in canola (*Brassica napus* L.) from hypocotyl segment explants. *Plant Cell Reports* 21: 599-604.

Cegielska-Taras, T., Pniewski, T. & Szała, L. (2008). Transformation of microspore-derived embryos of winter oilseed rape (*Brassica napus* L.) by using *Agrobacterium tumefaciens*. *Journal of Applied Genetics* 49: 343-347.

Cegielska-Taras, T., Tykarska, T., Szała, L., Kuraś, L. & Krzymański, J. (2002). Direct plant development from microspore-derived embryos of winter oilseed rape *Brassica napus* L. ssp. *oleifera* (DC.) Metzger. *Euphytica* 124: 341–347.

Chapel, M. & Glimelius, K. (1990). Temporary inhibition of cell wall synthesis improves the transient expression of the GUS gene in *Brassica napus* mesophyll protoplast. *Plant Cell Reports* 9: 105-108.

Chen, J.L. & Beversdorf, W.D. (1994). A combined use of microprojectile bombardment and DNA imbibition enhances transformation frequency of canola (*Brassica napus* L.). *Theoretical and Applied Genetics* 88: 187-192.

De Block, M., De Brower, D. & Tenning, P. (1989). Transformation of *Brassica napus* and *Brassica oleracea* using *Agrobacterium tumefaciens* and the expression of the *bar* and *neo* genes in the transgenic plants. *Plant Physiology* 91: 694-701.

Dill, G.M. (2005). Glyphosate-resistant crops: history, status and future. *Pest Management Science* 61: 219-224.

Dorman, M., Darla, N., Hayden, D., Puttick, D. & Quandt, J. (1998). Non-destructive screening of haploid embryos for glufosinate ammonium resistance four weeks after microspore transformation in *Brassica*. *Acta Horticulturae* 459: 191-197.

Friedt, W. & Snowdon, R. (2009). Oilseed rape. *in*: Handbook of Plant Breeding, vol. 4. Oil Crops, Vollman, J. & Rajcan, J. (eds.), Springer-Verlag, Dordrecht, Heidelberg, London, New York, pp. 91-126.

Friedt, W. & Zarhloul, K. (2005). Haploids in the improvement of Crucifers. *in*: Plamer, C.E., Keller, W.A., Kasha, K.J. (eds.), *Haploids in Crop Improvement II*, Springer-Verlag, Berlin, Heidelberg, pp. 191-213.

Fukuoka, H., Ogawa, T., Matsuoka, M., Ohkawa, Y. & Yano, H. (1998). Direct gene delivery into isolated microspores of rapeseed (*Brassica napus* L.) and the production of fertile transgenic plants. *Plant Cell Reports* 17: 323-328.

Genescan. (2011). Genetically Modified Canola. URL: *www.eurofins.de/genescan-en/other-services/gmo-information/gmo-canola*.

Grison, R., Grezes-Besset, B., Schneider, M., Lucante, N., Olsen, L., Leguay, J.-J. & Toppan, A. (1996). Field tolerance to fungal pathogens of *Brassica napus* constitutively expressing a chimeric chitinase gene. *Nature Biotechnology* 14: 643-646.

Golz, C., Köhler, F. & Scheider, O. (1990). Transfer of hygromycine resistance into *Brassica napus* using total DNA of a transgene *B. nigra* line. *Plant Molecular Biology* 15: 475-483.

Huang, B. (1992). Genetic manipulation of microspores and microspore-derived embryos. *In Vitro Cellular & Developmental Biology - Plant* 28P: 53-58.

ISAAA - International Service for the Acquisition of Agri-biotech Applications. (2011). Global Status of Commercialized Biotech/GM Crops: 2010. ISAAA Brief 42-2010: Executive Summary. URL: *www.isaaa.org/resources/publications/briefs/42/executivesummary*.

ISAAA - International Service for the Acquisition of Agri-biotech Applications. (2011a). Argentine Canola (*Brassica napus*) Events. URL: *www.isaaa.org/gmapprovaldatabase/cropevents*.

Jardinaud, M.F., Souvre, A. & Alibert, S.B. (1993). Transient *gus* gene expression in *Brassica napus* electroporated microspores. *Plant Science* 93: 177-184.

Jones-Velleneunes, E., Huang, B., Prudhomme, I., Sharon, B., Kemble, R., Hattori, J. & Miki, B. (1995). Assessment of microinjection for introducing DNA into uninuclear microspores of rapeseed. *Plant Cell Tissue Organ Culture* 40: 97-100.

Jonoubi, P., Mousavi, A., Majd, A., Salmanian, A.H., Jalali Javaran, M. & Daneshian, J. (2005). Efficient regeneration of *Brassica napus* L. hypocotyls and genetic transformation by *Agrobacterium tumefaciens*. *Biologia Plantarum* 49: 175-180.

Kazan, K., Curtis, M.D., Goulter, K.C. & Manners, J.M. (1997). *Agrobacterium tumefaciens*-mediated transformation of double haploid canola (*Brassica napus* L.) lines. *Australian Journal of Plant Physiology* 24: 97-102.

Kahrizi, D., Salmanian, A.H., Afshari, A., Moieni, A. & Mousavi, A. (2007). Simultaneous substitution of Gly96 to Ala and Ala183 to Thr in 5-enolpyruvylshikimate-3-phosphate synthase gene of *E. coli* (k12) and transformation of rapeseed (*Brassica napus* L.) in order to make tolerance to glyphosate. *Plant Cell Reports* 26: 95-104.

Maheshwari, P., Selvaraj, G. & Kovalchuk, I. (2011). Optimization of *Brassica napus* L. (canola) explant regeneration for genetic transformation. *New Biotechnology*, doi: 10.1016/j.nbt.2011.06.014.

McGloughlin, M.N. (2010). Modifying agricultural crops for improved nutrition. *New Biotechnology* 27: 494-504.

Moghaieb, R.E.A., El-Awady, M.A., El Mergawy, R.G., Youssef, S.S. & El-Sharkawy, A.M. (2006). A reproducible protocol for regeneration and transformation in canola (*Brassica napus* L.) *African Journal of Biotechnology* 5: 143-148.

Nehlin, L., Möllers, C., Bergman, P. & Glimelius K. (2000). Transient β-*gus* and *gfp* gene expression and viability analysis of microprojectile bombarded microspores of *Brassica napus* L. *Journal of Plant Physiology* 156: 175-183.

Oelck, M.M., Phan, C.V., Eckes, P., Donn, G., Rakow, G. & Keller, W.A. (1991). Field resistance of canola transformants (*Brassica napus* L.) to Ignite™ (phosphinotricin). *Proceedings of the 8th International Rapeseed Congress*, GCIRC, Saskatoon, Canada, pp. 292-297.

Pechan, P.M. (1989). Successful cocultivation of *Brassica napus* microspores and proembryos with *Agrobacterium*. *Plant Cell Reports* 8: 387-390.

Poulsen, G.B. (1996). Genetic transformation of *Brassica*. *Plant Breeding* 115: 209-225.

Senior, I.J. & Bavage, A.D. (2003). Comparison of genetically modified and conventionally derived herbicide tolerance in oilseed rape: A case study. *Euphytica* 132: 217-226.

Sharma, K., Bhatnagar-Mathur, P. & Thorpe T.A. (2005). Genetic transformation technology: Status and problems. *In Vitro Cellular & Developmental Biology - Plant* 41: 102-112.

Snowdon, R., Lüchs, W. & Friedt, W. (2007). *Brassica* oilseeds. *in*: Singh, R.J. (ed.), *Genetic resources, chromosome engineering, and crop improvement*. CRC Press, Boca Raton, pp. 195-230.

Swanson, F.B. & Erickson, I.R. (1989). Haploid transformation of *Brassica napus* using an octopine-producing strain of *Agrobacterium tumefaciens*. *Theoretical and Applied Genetics* 78: 831-835.

Takahata, Y., Fukuoka, H. & Wakui, K. (2005). Utilization of microspore-derived embryos. *in*: Plamer, C.E., Keller, W.A., Kasha, K.J. (eds.), *Haploids in Crop Improvement II*, Springer-Verlag, Berlin, Heidelberg, pp. 153–169.

Troczyńska, J., Drozdowska, L. & Cegielska-Taras, T. (2003). Transformation of microspore-derived embryos to study the myrosinase-glucosinolate system in *Brassica napus* L. *Proceedings of the 11th International Rapeseed Congress*, GCIRC, Copenhagen, Denmark, pp. 175-177.

Vollmann, J. & Rajcan, I. (2009). Oil Crops Breeding and Genetics. *in*: *Handbook of Plant Breeding*, vol. 4. Oil Crops, Vollman, J. & Rajcan, J. (eds.), Springer-Verlag, Dordrecht, Heidelberg, London, New York, pp. 1-30.

Wang, J., Chen, Z., Du, J., Sun, Y. & Liang, A. (2005). Novel insect resistance in *Brassica napus* developed by transformation of chitinase and scorpion toxin genes. *Plant Cell Reports* 24: 549-555.

Wang, W.C., Menon, G. & Hansen G. (2003). Development of a novel *Agrobacterium*-mediated transformation method to recover transgenic *Brassica napus* plants. *Plant Cell Reports* 22: 274-281.

Wang, Y.P., Sonntag, K., Rudloff, E. & Han, J. (2005). Production of fertile transgenic *Brassica napus* by *Agrobacterium*-mediated transformation of protoplasts. *Plant Breeding* 124: 1-4.

Zhong, G.Y. (2001). Genetic issues and pitfalls in transgenic plant breeding. *Euphytica* 118: 137–144.

7

Use of Tebuthiuron to Restore Sand Shinnery Oak Grasslands of the Southern High Plains

David A. Haukos[1]

U.S. Fish and Wildlife Service, Department of Natural Resources Management, Texas Tech University, Lubbock, Texas, USA

1. Introduction

The Southern High Plains of northwest Texas and eastern New Mexico represents the extreme southwest subdivision of the Great Plains. This 130,000 km² plateau represents a "remnant of the Rocky Mountain piedmont alluvial plain" with borders abruptly demarcated by the Canadian river to the north, Pecos river to the west, and the dramatic Caprock Escarpment to the east (Holliday, 1990; Reeves, 1965; Fig. 1). The southern border is relatively undefined as a gradual merging into the Edwards Plateau. As one of the last regions to be permanently settled in the conterminous United States, the semi-arid Southern High Plains was frequently described as a desolate, never-ending featureless landscape unsuitable for human occupation and agriculture (Wester, 2007). The primary factor underlying the fear of humans traveling across the Southern High Plains was the lack of reliable surface water. However, the discovery of southern extent of the massive underlying Ogallala Aquifer combined with the development of the deep-well technology in the 1940s to mine large quantities of water lead to the conversion of the Southern High Plains to one of the most agriculturally impacted regions of the world (Smith, 2003; Wester, 2007).

Geologically, the Southern High Plains is comprised of two main formations. The Quaternary Blackwater Draw Formation formed by multiple episodes of eolian sheet deposition during the past 1.4 million years (Holliday, 1990). Beneath the Blackwater Draw Formation is the Miocene-Pliocene Ogallala Formation that was created between about 11 million to 1.4 million years ago (Holliday, 1990). Holliday (1990) suggested that the Southern High Plains has likely been a grassland or savanna grassland for about 11 million years based on evidence for an arid to a semi-arid or sub-humid environment. Ecologically, the Southern High Plains is currently considered a short-grass prairie dominated by buffalo grass (*Bouteloua dactyloides*) and blue grama (*Bouteloua gracilis*) prior to settlement (Blair, 1950; Kuchler, 1970). This community has likely been present for approximately 10,000 years (Axelrod, 1985; MacGinitie, 1962). The natural ecological drivers of this system include fire (natural and prescribed), large herbivores (e.g., modern bison [*Bison bison*]), and extreme

---

[1] Current Address: U.S. Geological Survey, Kansas Cooperative Fish and Wildlife Research Unit, Kansas State University, Manhattan, Kansas 66506

unpredictable environmental conditions (e.g., extended droughts, short periods of intensive precipitation, temperatures ranging from -33 to 44 C) (Stebbins, 1981; Wester, 2007; Wright & Bailey, 1982).

The featureless landscape of the Southern High Plains is relieved only by Holocene dune fields (formed during droughts of the Altithermal period) along its southwestern borders (Holliday, 1989), >20,000 ephemeral playa wetlands scattered across the landscape (Bolen et al., 1989; Haukos & Smith, 1994; Smith, 2003), approximately 40 historically spring-fed saline lakes, and several currently dry, but historically spring-fed tributaries (i.e., "draws") of the Colorado, Brazos, and Red Rivers (Holliday, 1990). The deep sandy soils associated with the dune fields, found primarily in the southwest portion of the Southern High Plains, along the Texas – New Mexico border, form a unique and distinctive ecosystem about which little is known ecologically.

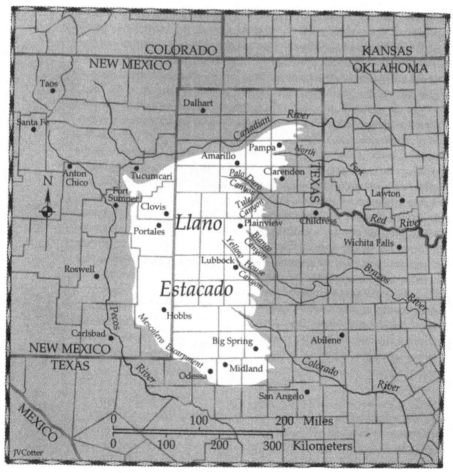

Fig. 1. The Southern High Plains of northwest Texas and eastern New Mexico (from Morris, 1997).

These sandy soils primarily developed during the warmer, drier conditions between 4,000 – 7,000 years ago following the episodic deposition of sheet sands between 8,000 and 11,000 years ago and represent cumulative effects of drought, eolian deposition, and distribution of ancient waterways (Gile, 1979; Holliday, 2001). Eolian sands in dune fields and sand sheets cover >10,000 km^2, about 10% of the Southern High Plains, with approximately 5,800 km^2 existing as dune fields (Holliday, 2001). The Blackwater Draw Formation was locally buried by the deposition of sand during the periods of deposition (Holliday, 1989). Until the early 1900s, active dunes were present in the region (Muhs and Holliday, 1994). The topography of these areas differ dramatically from the surrounding short-grass prairie, which has little relief whereas the sandy soils support parabolic and coppice dunes, blow-outs, and sand sheets that creates a varied and heterogeneous landscape (Holliday, 2001). Although not yet quantified, a wide variation of micro-climates occurs within the landscape, which contrasts with the relatively few micro-climates of the short-grass prairie. According to Muhs and Holliday (2001) the dunes of the Southern High Plains are comprised of the most mineralogically mature sands of the Great Plains. Loss of vegetation cover allowing erosion of the Blackwater Draw Formation destabilizes the dune system and adds new sand to the system (Muhs and Hollida,y 2001).

Although there are three identified west to east trending dune fields in the southwestern Southern High Plains – Muleshoe dunes, Lea-Yoakum dunes, and Andrews dunes – the focus of this chapter is on vegetation of the Muleshoe and Lea-Yoakum dunes, which are the northernmost extensive dune fields with associated sand sheets (Holliday, 2001). Relative activity of dunes ranges from least active in the northern Muleshoe dunes to most active in the Andrews dunes, the driest, warmest region (Muhs and Holliday, 2001). The Muleshoe dunes are higher, more vegetated, and relatively stable than the Lea-Yoakum dunes (Holliday, 2001), but support similar plant communities. The primary soil series is Brownfield-Tivoli fine sand (Neuman 1964), but additional soil types (mostly Entisols, Alfisols, Aridsols, and Mollisols) that support similar plant communities can be found (Garrison et al., 1977; Pettit, 1986, 1994). The sandy soils historically contained numerous springs and perched water tables hidden among the dunes (Brune, 2002; Marcy, 1850; Smith, 1985,). With the advent of irrigation depleting the Ogallala Aquifer, these springs have dried and the location of most remains unknown.

These dune fields and sandy soils have always supported taller grasses and unique woody shrubs than the short-grass prairie of surrounding clay and sandy loams (Shantz and Zon, 1924). A unique ecosystem of sand shinnery oak (*Quercus harvardii*) and mid to tall grasses (e.g., *Andropogon hallii, Andropogon gerardii, Schizachyrium scoparium*, and *Sporobolus cryptandrus*) evolved in response to the soil and environmental conditions and represents a relic area of distinctive biodiversity relative to the short-grass prairie. As clay content in the soil increases, cover of sand shinnery oak decreases (Pettit, 1978) as the species grows best on sites with an almost pure sand cover (Small, 1975). Estimated coverage of sand shinnery oak presettlement is 5-25% (Conner et al., 1974; Hodson et al., 1980). The species has been present for at least 3000 years based on pollen profile (Gross and Dick Peddie, 1979; Hafsten, 1961,). There is a relatively high diversity of forbs in the community as well (Peterson and Boyd, 1998). Ecological drivers for this system are the same as for the short-grass prairie, but grazing and fire were likely more infrequent historically. Bison were apparently attracted to water but the deep sands were difficult to traverse by such a large mammal so it is unlikely that large herds frequently grazed the interior of these habitats. Fire does occur and structures the community, but likely not as recurrent as in the short-grass prairie because of the need to accumulate larger fuel loads to carry a fire across the patchy vegetation of the sandy soils. Pettit (1979) stated "These lands are perhaps the most fragile of all ecosystems on the Southern High Plains of Texas and the landowner cannot afford to abuse them."

Typically not greater than 1 m tall, sand shinnery oak is the plant species most directly associated with this community. Although occurring in other regions of the United States, sand shinnery oak likely has the greatest ecological influence in the sandy soils of the southwestern Southern High Plains (Peterson & Boyd, 1998; Fig. 2). The natural form of sand shinnery oak in this region is that of a low shrub with up to 100 or greater short, aerial shoots from a massive underground stem and root system (Peterson & Boyd, 1998; Pettit, 1986). The underground root system is the primary reproductive structure for sand shinnery oak.

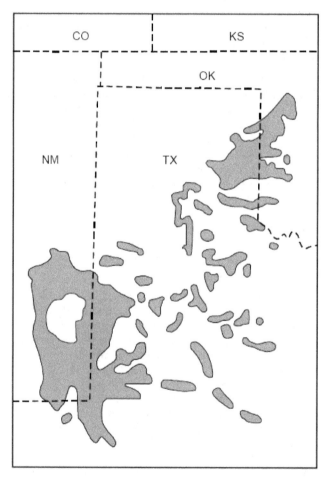

Fig. 2. Distribution of sand shinnery in New Mexico, Texas, and Oklahoma (Peterson & Boyd, 1989).

The stable existence of sand shinnery oak for thousands of years (Beckett, 1976; Gross and Dick-Peddie, 1979) and lack of spread by acorns refutes any claims that sand shinnery oak has increased in abundance and range in the past century. However, historically, in the absence of grazing, grasses frequently overtop and obscure sand shinnery oak (Brown, 1982; Duck and Fletcher, 1944). Furthermore, the roots of sand shinnery oak are also the primary soil stabilization structure. Loss of sand shinnery oak typically results in severe erosion of the sandy soils without the presence of vegetation (Parks, 1937; Moldenhauer et al., 1958).

Historical estimates of area of sand shinnery oak greatly vary, but Peterson & Boyd (1998) estimated that the species covered 405,000 ha in Oklahoma, 607,000 ha in New Mexico, and 1.4 million ha in Texas. The species range has not apparently expanded since the mid-1800s (Gross and Dick-Peddie, 1979; McIlvain, 1954; Peterson & Boyd, 1998). By 1972, approximately 500,000 ha had been converted to cropland or pastureland in Texas (Deering & Pettit, 1972). Current estimates of the area of sand shinnery oak are unreliable; but since 1972, the development of center pivot irrigation systems and advances in herbicide technology have resulted in a considerable reduction of sand shinnery oak habitats in the Southern High Plains (Bailey & Painter, 1994; Dhillion et al., 1994).

Sand shinnery oak has a unique life history. The species is deciduous with bud development in early to late March (Pettit, 1986), leaves open during April and May (Boo & Pettit, 1975; Pettit, 1975), flowering occurs in April and May (Rowell, 1967), and leaf drop normally happens in early November (Boo & Pettit, 1975). Acorns mature in July (Peterson & Boyd, 1998). Ninety-five percent of root growth is during July through September (Zhang, 1996). Most grasses and forbs in sand shinnery oak communities of the southwest Southern High Plains develop following precipitation events from May-July (Galbraith, 1983; Pettit, 1979).

Sand shinnery oak has uniquely evolved in the semi-arid environment of the Southern High Plains to efficiently gather and store water. Up to 50% of plant mass can be water during periods of precipitation (Pettit, 1986). However, during multi-year droughts, sand shinnery oak may not leaf out or may lose its leaves during the growing season to reduce water loss (Jones and Pettit, 1980; Jones and Pettit, 1984). However, the long-term effects of drought are more evident on grasses than sand shinnery oak because of the ability of the oak to store water and carbohydrates (Galbraith, 1983), which permits a relatively rapid response to alleviation of drought conditions compared to grasses that may not return to predrought production until the following growing season.

During spring bud and leaf emergence, catkins, buds, and leaves have a high phenolic content that can be poisonous to livestock (Allison, 1994; Dollahite, 1961; Peterson & Boyd, 1998). Therefore, sand shinnery oak is avoided by livestock at this time or, in most situations, livestock are removed from pastures that are predominately sand shinnery oak. Ingestion of sand shinnery oak buds and leaves during this time may lead to general malaise, reduced conception rates, lower weight gains, and death (Jones & Pettit, 1984). However, sand shinnery oak can be a significant portion of livestock diet especially during late summer and fall (Dayton, 1931; Plumb and Pettit, 1983; Roebuck, 1982).

Sand shinnery oak rarely reproduces through sprouting from produced acorns (Dhillion et al., 1994; Pettit, 1977; Wiedeman, 1960). This is likely the result of insufficient soil moisture because acorns are viable (Pettit, 1977; Pettit,1986). However, acorn crops occur on average in 3 of 10 years (Pettit, 1986). Therefore, reproduction is primarily sprouting from rhizomes

(Pettit, 1977). Fire will top-kill sand shinnery oak, but vigorous resprouting usually occurs within a year, depending on soil moisture conditions (Slosser et al., 1985). Of great concern, however, is a lack of moisture following a prescribed or natural fire can result in significant wind erosion of the sandy soils (Zobeck et al., 1989).

Sullivan & Pettit (1977) reported that sand shinnery oak was most productive in younger soils. In the southwestern Southern High Plains, monocultures of sand shinnery oak produce 3,300 kg/ha of air-dried above-ground forage (Lenfesty, 1983). Sears et al. (1986) reported growing-season above-ground biomass of sand shinnery to be 1,821 kg/ha but a below-ground growing season biomass of 19,841 kg/ha. This corresponds with the estimate by Pettit (1986) that the ratio of underground to above-ground tissues was 10:1 to 16:1, which is likely greater than that of any other North American shrub. Depth of roots is frequently >5 m but above-ground shoots are rarely exceed 0.75 m tall in the southwestern Southern High Plains (McIlvain, 1954). The below-ground system of a single plant can be 3 – 10 m or more in diameter (Muller, 1951). Density of above-ground shoots can be 30 – 75 individuals/m^2 (Jones, 1982; Zhang, 1996). Mayes (1994) reported a single clone covering > 7,000 m^2. The maximum extent and age of individual sand shinnery oak plants are unknown, but age of below-ground structures is likely measured in centuries (Peterson & Boyd, 1998). Individual above-ground shoots usually have a lifespan of 11-15 years (Muller, 1951; Pettit, 1986). Sand shinnery oak can spread at a rate of 1 m per 5 years in fields where the species was previously removed (Sikes & Pettit, 1980), which reduces the likelihood of the species being considered invasive.

Understanding of the need for sand shinnery oak by vertebrates is mixed. Many species depend on the habitats supported by the sand-shinnery oak – grassland complex; including lesser prairie-chicken (*Tympanuchus pallidicinctus*) and dune sagebrush lizard (aka: sand dune lizard; *Sceloporus arenicolus*), which are candidates for listing as endangered or threatened species under the Endangered Species Act of the United States. The endemic dunes sagebrush lizard is restricted to sand shinnery oak habitats (Degenhardt & Jones, 1972; Degenhardt et al., 1996). The species is predominately found in areas of open sand (i.e., blow outs) but uses sand shinnery oak nearly exclusively for forage, thermal cover, and escape habitat. Due to its status of under review for listing as an endangered species, eradication of sand shinnery oak is restricted on public lands containing dune sagebrush lizard habitat (USDI BLM, 1997).

Approximately 100 avian species, including numerous species of conservation concern, such as Cassin's sparrow (*Aimophila cassinii*), use sand shinnery oak habitats for nesting, migration, and wintering. Scale quail (*Callipepla squamata*), northern bobwhite (*Colinus virginianus*), and mourning dove (*Zenaidura macroura*) use sand shinnery oak habitats during the nesting season. Approximately additional 20 species of songbirds nest in sand shinnery oak and 80 species use the habitat at some point during the year (Peterson & Boyd 1998). Raptors present in sand shinnery oak habitats include Mississippi kite (*Ictinia mississippiensis*), northern harrier (*Circus cyaneus*) Harris' hawk (*Parabuteo unicinctus*), Swainson's hawk (*Buteo swainsoni*), sharp-shinned hawk (*Accipiter stiatus*), Cooper's hawk (*Accipiter cooperii*), rough-legged hawk (*Buteo regalis*), ferruginous hawk (*Buteo lagopus*), golden eagle (*Aquila chrysaetos*), prairie falcon (*Falco mexicanus*), barn owl (*Tyto alba*), and great-horned owl (*Bubo virginianus*) (Peterson & Boyd, 1998). NMPIF (2007) has identified eight bird priority species of concern using sand shinnery oak communities.

Mule deer (*Odocoileus hemionus*) feed on sand shinnery oak acorns, buds, and leaves (Gray et al., 1978; Krysl et al., 1980; Ligon, 1927). White-tailed deer (*Odocoileus virginianus*) are also found in sand shinnery oak habitats (Ligon, 1927; Raught, 1967). Pronghorn (*Antilocapra americana*) are frequently found in the sandhills using sand shinnery oak (Roebuck & Simpson, 1982). Black-tailed jackrabbits (*Lepus californicus*) and desert cottontails (*Sylvilagus auduboni*) are commonly found in sand shinnery oak habitats (Peterson & Boyd, 1998). Sixteen small mammal species have been reported in areas of sand shinnery oak, but none are endemic (Peterson & Boyd, 1998). The most common mammalian predators are coyote (*Canis latrans*), bobcat (*Lynx rufus*), badger (*Taxidea taxis*), striped skunk (*Mephitis mephitis*), and swift fox (*Vulpes velox*) (Peterson & Boyd, 1998). Reptiles are represented by western box turtle (*Terrapene ornate*) and 25 species of snakes including plains hognose snake (*Heterodon nasicus*), coachwhip (*Masticophis flagellum*), bull snake (*Pituophis melanoleucus*), massasauga (*Sistrurus catenatus*), and prairie rattlesnake (*Crotalus viridis*) (Peterson & Boyd 1998). In addition to the dunes sagebrush lizard, there are 9 lizard species that use sand shinnery oak habitats in southwestern Southern High Plains including prairie lined racerunner (*Cnemidophorus sexlineatus*), western whiptail (*C. tigris*), Great Plains skink (*Eumeces obsoletus*), leopard lizard (*Gambelia wislizenii*), lesser earless lizard (*Holbrookia maculate*), Texas horned lizard (*Phrynosoma cornutum*), and prairie lizard (*Sceloporus undulates*) (Degenhardt and Jones, 1972; Degenhardt et al., 1996; Gorum et al., 1995; Peterson & Boyd, 1998; Wolfe, 1978).

Invertebrates are an important component of sand shinnery oak with at least 23 families represented (Haukos & McDaniel, 2011). Annual biomass of invertebrates dramatically fluctuates, which does not appear to be related to annual precipitation (Haukos & McDaniel, 2011). Many species depend on invertebrates to persist in sand shinnery oak habitats, but little is known about the ecology of invertebrates in these habitats of the Southern High Plains.

The most widely identified vertebrate species of sand shinnery oak in the southwestern Southern High Plains is the lesser prairie-chicken. Although opinions vary regarding the importance of sand shinnery oak for lesser prairie-chickens, the accumulated evidence suggests that the habitats supported by sandy soils, including sand shinnery oak, comprise the core of the historic and current population in the region. Due to a >90% decline in numbers and range (Crawford, 1980; Taylor & Guthery, 1980), the species has been a candidate for listing as an endangered species since the late 1990s under the United States federal Endangered Species Act.

Lesser prairie-chickens consume acorns, galls, catkins, and vegetation of sand shinnery oak and insects associated with sand shinnery oak (Crawford & Bolen, 1976; Davis et al., 1981; Doerr and Guthery, 1983; Riley et al., 1993). Lesser prairie-chickens will lek, nest, and raise broods in sand shinnery oak habitats (Haukos & Smith 1989; Riley et al., 1993; Sell, 1979). Nests are usually constructed in residual grasses but frequently surrounded by sand shinnery oak as protective cover (Haukos & Smith, 1989; Riley et al., 1992; Sell, 1979). Sand shinnery oak provides thermal cover, escape cover, and roosting cover for lesser prairie-chickens (Copelin, 1963; Crawford & Bolen, 1976; Davis et al., 1981; Riley et al., 1993; Sell, 1979).

Crawford (1974) reported that sand shinnery oak comprised 15% of volume of fall foods. Davis et al. (1979) reported that sand shinnery oak acorns comprised 39 and 69% of fall and winter diets, respectively, of lesser prairie-chickens. They also noted that sand shinnery oak leaf galls (produced by a parasitic wasp) and catkins were also important foods.

From a management perspective, land ownership of sand shinnery oak landscapes on the Southern High Plains has a tremendous influence on management strategies. In Texas, the vast majority of sand shinnery oak habitat is on private land that supports agricultural activities or oil/gas development. Whereas, in New Mexico, there is considerable area of sand shinnery oak habitats held in public trust (~700,000 ha) that is managed by a variety of federal, state, and local governments in addition to nongovernment organizations (Peterson & Boyd, 1998). The sand shinnery oak habitats on public lands in New Mexico are managed for multiple uses including livestock grazing, oil and natural gas development, and hunting.

Although with some debate regarding the cause, current plant assemblages in sand shinnery oak – grass communities tend to be dominated by sand shinnery oak at the expense of grass coverage. Recent estimates of ground cover of sand shinnery oak are 80-90% (Biondini et al., 1986; Dhillion et al., 1994; Pettit, 1994; Plumb, 1984). Frequently, sand shinnery oak is categorized as increasing or invader under grazing pressure comparing to decreasing grass and forb component (Herbel, 1979; Herndon, 1981; Lenfesty, 1983). However, there is little evidence that sand shinnery oak invades or increases in absolute density or abundance in overgrazed grassland despite the perception of development of a monoculture (Dickerson, 1985; Holland, 1994); but rather, when given a competitive advantage due to grazing pressure as an effective water gatherer (Galbraith, 1983; Pettit, 1986; Sullivan, 1980; Zhang, 1996), will reduce or eliminate associated species due to the effects of shading and moisture competition. Thus, overgrazing and suppression of fire has reduced or eliminated the herbaceous (both grasses and forbs) species associated with sand shinnery oak.

Most land managers consider this condition undesirable and noxious, with considerable effort expended to eradicate sand shinnery oak with varying degrees of success since the early 1970s. Restoration of the plant composition to historical proportions requires retarding growth of sand shinnery oak to allow development and growth of herbaceous species to a point of a stable community, which may occur when mid and tall grasses are permitted to fully develop. For example, when tall grasses overtop sand shinnery oak, clones are reduced in vigor and density (Frary, 1957; Muller, 1951). Indeed, Moldenhauer et al. (1958) stated that under natural conditions, grasses were dominant and outcompeted sand shinnery oak.

Clearing of sand shinnery oak and planting crops is effective in reducing oak but is very expensive, removes critical habitat, and results in short-term productivity that requires significant nutrient inputs within a few years of clearing (Peterson & Boyd 1998). Additionally, complete removal of sand shinnery oak for croplands results in short-term success with the high potential for wind erosion of topsoil (Lotspeich & Everhart, 1962). Control of sand shinnery oak using prescribed fire usually is a short-term benefit because of vigorous resprouting within 2-3 growing seasons following the fire (McIlvain, 1954).

The most common method used to reduce or eradicate sand shinnery oak is application of herbicides. Initial efforts were application of phenoxy herbicides including 2, 4-D and 2, 4, 5-T; benzoic acids (i.e., dicamba); and picolenic acid (i.e., picloram) as liquids for absorption through foliage (Peterson & Boyd 1998). Typically, use of these herbicides at >1 kg/ha active ingredient (ai) resulted in the top-killing of 85-95% of sand shinnery oak (Greer et al., 1968; Pettit, 1977). To avoid removal of associated grasses and forbs that occurs at high rates of application (e.g., 3 – 8 kg ai/ha) necessary to kill sand shinnery oak with a single application, annual spraying for 2-3 years at lower rates has been recommended (Pettit, 1976; Pettit, 1977). Grass production can be doubled to quadrupled for a few years following herbicide application prior to exhibiting a decline (Greer et al., 1968; McIllvain & Armstrong, 1959). Thus, initial use of herbicides to control sand shinnery oak was exceptionally expensive relative to the duration of the effect.

In 1974, a pellet form of tebuthiuron (trade name of Spike® or Graslan®) applied to the soil and absorbed by roots began being used on sand shinnery oak (Peterson & Boyd, 1998). Tebuthiuron causes repeated defoliation of oak, which causes death within 2-3 years following application (Jones & Pettit, 1984; Peterson & Boyd, 1998). At 3 kg ai /ha, sand shinnery oak is killed, but many nontarget grasses and forbs are also killed (Pettit, 1979). Furthermore, at high application rates, much of the increase in annual production is by annual or undesirable grasses (Jacoby et al., 1983; Plumb, 1984). By the mid-1990s, at least 130,000 and 40,000 ha of sand shinnery oak had been treated with tebuthiuron in Texas and New Mexico, respectively (Johnson & Ethridge, 1996). The advantages of tebuthiuron include its relative nontoxicity to nontarget species (Emmerich, 1985), its effectiveness after only one application (Scifres et al., 1981), and elimination of overspray that is a characteristic of liquid herbicides (Scifres et al., 1981).

There is considerable variation of the magnitude of sand shinnery oak kill and resultant grass response to use of tebuthiuron. At the relatively low rate of 0.4 kg/ha, cover of sand shinnery oak was reduced by 95% and grass yield increased 2.5 times controls after 3 years (Jones & Pettit, 1984). At rates of 0.6 – 1.0 kg/ha, oak is usually killed (Peterson & Boyd, 1998). The maximum grass yield of 4 times the control was found at a rate of 0.8 kg/ha (Jones & Pettit, 1984). Doerr (1980) found that rates of tebuthiuron from 0.2 – 1.0 kg/ha increased grass coverage from 88-130% and density of bunchgrasses from 12-32% by decreasing density of sand shinnery oak at least 84%. Forb densities and grass production were decreased in plots with 0.8 and 1.0 kg/ha. Seed production was increased in plots with application rates of 0.2, 0.4, and 0.6 kg/ha. Depending on rate of application, treatment with tebuthiuron tends to decrease vertical screening immediately after application as the shinnery oak dies, but as bunchgrasses recover, vertical screening in treated plots may surpass that in untreated plots (Doerr & Guthery, 1983). Likewise, canopy cover eventually can be greater in tebuthiuron treated plots than in untreated plots (Doerr & Guthery, 1983; Jones, 1982). Doerr (1980) recommended an application rate of 0.4 kg/ha because of the increased grass response in the first year following treatment relative to greater rates. Scifres & Mutz (1978) reported that most forb species were killed at application rates of 2.0 and 3.0 kg/ha. A 25% kill of sand shinnery oak prevents acorn production for up to 2 years, which is a reduction of an important component of forage for lesser prairie-chickens and other wildlife (Peterson & Boyd, 1998). Crude protein in grasses increased by 28% for one growing season post tebuthiuron application (Biondini et al., 1986). Forbs generally increase in diversity and production >2 years after application (Doerr & Guthery, 1983; Jacoby et al., 1983; Jones & Pettit, 1982; Olawsky & Smith, 1991). Sears et al. (1986) found a 17% decrease in sand shinnery oak and 266% increase in herbaceous vegetation within three years of treatment. Six years after treatment, above-ground biomass had decreased due to a 41% decrease in sand shinnery oak, 32% decrease in litter, and 161% increase in herbaceous vegetation (Sears et al., 1986).

Without government assistance it is unlikely that treating sand shinnery oak in New Mexico (lowest precipitation and grass response) would be economical for livestock producers unless changes are made to grazing systems to ensure that grasses persist as a component of the plant assemblage (Peterson & Boyd, 1998). From an economic perspective, long-term cost-effectiveness of treating sand shinnery oak with tebuthiuron varies due to precipitation, beef prices, herbicide cost, and grazing management following treatment (Etheridge et al., 1987a; Etheridge et al., 1987b).

In contrast to plant response, there have been few investigations into faunal response to the use of tebuthiuron to control sand shinnery oak. The vast majority of studies focus on lesser prairie-chickens. At high rates (>1.5 kg ai/ha) application of tebuthiuron has resulted in preferred use by lesser prairie-chickens of untreated areas during nesting and brood rearing, indicating selection for habitats containing some cover by sand shinnery oak, but the extent of this response may be somewhat confounded by grazing pressure (Haukos & Smith, 1989). Sell (1979) found 75% of lesser prairie-chicken nests in sand shinnery oak or sand sagebrush. Haukos & Smith (1989) reported that 80% of nesting lesser prairie-chickens nested in untreated sand shinnery oak. Taylor (1978) reported that lesser prairie-chickens preferred habitats dominated by sand shinnery oak with a grass component during fall and winter. Donaldson (1966; 1969) reported an increase in lesser prairie-chickens in sand shinnery oak treated areas compared to untreated but also indicated that sand shinnery oak was the plant most commonly used even in treated plots. Further, he proposed supplemental winter feeding to make up for loss of acorns due to treatment. Olawsky (1987) reported that acorns were the major food of lesser prairie-chickens in untreated areas, but absent in treated areas, which resulted in lower lipid levels (i.e., lower body condition).

There is little conclusive evidence from Texas or New Mexico that treatment of sand shinnery oak with tebuthiuron benefits lesser prairie-chickens (Pederson & Boyd, 1999). There was no statistical difference in density of lesser prairie-chickens between treated and untreated sand shinnery oak during summer (0.51 and 0.41 birds/ha in treated and untreated plots, respectively) and winter (0.53 and 0.35 birds/ha in treated and untreated plots, respectively) (Olawsky et al., 1988). Martin (1990) reported 86% fewer lesser prairie-chickens in treated versus untreated sites, but indicated detection was difficult in grass pastures. Haukos & Smith (1989) reported that lesser prairie-chickens preferred to nest in untreated sites, likely due to intense grazing pressure in treated sites. Wide-spread eradication of sand shinnery oak is thought to be exceptionally detrimental to lesser prairie-chickens (Davis et al., 1979; Doerr & Guthery, 1980; Olawsky & Smith, 1991; Riley et al., 1993; Sell, 1979; Taylor & Guthery, 1980,). Johnson et al. (2004) observed lesser prairie-chicken hens selecting sand shinnery oak dominated rangelands not treated with herbicide, for nest sites, significantly more than herbicide treated rangelands. Johnson et al. (2004) also reported greater density of shrubs within a 3-m radius surrounding the lesser prairie-chicken nest sites. In New Mexico, Bell et al. (2010) most often located lesser prairie-chicken broods in dense sand shinnery oak areas. Lesser prairie-chicken survival was greatest in habitats with shrub density ≥20% (Patten et al., 2005).

Reduction in acorn production due to tebuthiuron treatment is considered detrimental for deer (Bryant & Demarais, 1992; Bryant & Morrison, 1985). Bednarz (1987) reported that lagomorphs populations were reduced following eradication of sand shinnery oak with 2-3 years following spraying necessary for recovery. At a rate of 0.56 kg/ha, Colbert (1986) found that populations of small mammals were not affected by tebuthiuron treatment. Doerr & Guthery (1980) found rodent populations to be 41% greater on untreated plots compared to plots where 75% of sand shinnery oak was removed. Ord's kangaroo rat seems to be the only small mammal that responds positively to tebuthiuron control of sand shinnery oak (Colbert, 1986; Fischer, 1985; Willig et al., 1993). With the exception of the Great Plains skink, lizards were more commonly found in untreated than treated sand shinnery oak habitats in New Mexico (Gorum et al., 1995; Snell et al., 1997). Martin (1990) reported that reduction of sand shinnery oak by 90% did not change number of birds and avian richness, but Cassin's sparrows may have increased slightly on treated plots and lesser prairie-chickens decreased on treated plots. However, meadowlark populations may be double in treated sand shinnery oak versus untreated (Olawsky et al., 1987).

Recommendations for use of herbicides to reduce sand shinnery oak and increase herbaceous production for the purpose of increasing weight gain of livestock are (1) shin-oak should not be treated in drought years, which are difficult to predict, (2) areas of large dunes should not be treated due to erosion potential, (3) areas with little cover of existing perennial grass species should not be treated prior to 1-2 seasons of grazing exclusion, and (4) treated areas should not be burned or grazed during the growing season for at least 2 years following treatment (Doerr & Guthery, 1980; McIlvain and Armstrong, 1959). Doerr (1980) suggested that two rates be used to manage for lesser prairie-chickens (0.2 and 0.6 kg/ha) avoiding treatment of sand dunes and following light or no grazing pressure to ensure residual vegetation cover to reduce threat of wind erosion following treatment. Davis et al. (1974) recommended control of sand shinnery oak to be in strips to benefit birds. Unfortunately, these recommendations are rarely followed in practice resulting in significant reduction of dune topography, reduced sustained grass response to treatment, invasion of undesirable and difficult to eradicate plant species, and increased potential for major erosion events (Thurmond et al., 1986; Zobeck et al., 1989). Not only is grazing management critical following application of tebuthiuron to ensure sustain production of desirable grasses, but lesser prairie-chickens, and perhaps other species, respond to intensive grazing following treatment by using untreated sand shinnery oak. Overgrazing of grasses following treatment of sand shinnery oak can result in the conversion to dominance by sand sagebrush, which is much more difficult to control (pers. observ.).

Recently, throughout the southwestern Southern High Plains, there has been interest in restoring the vegetation communities to a more historic grass/shrub balance. Since approximately 2000, rates of wide-spread tebuthiuron application have been reduced (e.g., <1.0 kg/ha) with the avoidance of treating dune areas in an effort to temporarily reduce extent of sand shinnery oak and competitively release grass species to restore the historical balance of sand shinnery oak and grasses (Smythe & Haukos, 2010). Data on community response to reduced tebuthiuron application rates (0.60 kg ai/ha and dune avoidance) to restore the historical oak-grass under moderate grazing pressure initiated three years following herbicide application were collected as part of a 10-year study (Smythe & Haukos, 2009; Smythe & Haukos, 2010). Tebuthiuron was applied at to 532 ha of private land in 2000, which was adjacent to 518 ha of untreated land owned by the state of New Mexico representing the extant shinnery oak-grassland community. This rate of application rate was approximately 50% of previously recommended rates for the area to ensure that sand shinnery oak was not completely eliminated from the community. The control area had not been grazed for 7 years before the study began; tebuthiuron-treated areas had not been grazed for 2 years pre-tebuthiuron treatment and 3 years post-tebuthiuron treatment. Grazing treatment was a short-duration system in which plots were grazed once during the dormant season (January and February) and once during the growing season (July). Stocking rate was calculated each season based on measured forage production and designed to take 25% of available herbaceous material per season.

There was a 6.5-fold increase in herbaceous plant production and a 29-fold difference in grass seed production on treated versus untreated areas (Smythe, 2006). The treated, ungrazed plots consistently had the greatest visual obscurity, whereas untreated, grazed plots had the lowest. Treated plots had greater visual obscurity at about 0.5 m and greater maximum height of vegetation. Vegetation was tallest in tebuthiuron-treated plots. Overhead cover did not differ among treatments. However, differences in vertical density among treatments occurred only at heights >40 cm.

Nesting grassland birds did not exhibit selection among nest sites based on vertical density, nor did vertical density affect hatching success (Smythe & Haukos, 2009). At lower levels of vegetation, those most important for concealment of nests, there was no difference in vertical density among treatments and no need for birds to select for nest sites among treatments. Average height of shinnery oak on the study site was 46.4 cm. This indicates that at lower vegetational strata, untreated shinnery oak provides similar vertical screening as the predominantly little bluestem communities that replace them after treatment with tebuthiuron (Smythe & Haukos, 2009).

Application of tebuthiuron at 0.60 kg/ha to restore sand shinnery oak communities in New Mexico, resulted in increased density of grassland birds (Smythe & Haukos, 2010). Treated sand shinnery oak plots restored to a grass/shrub mix supported a greater density of spring migrants and breeding birds than untreated plots. Migratory birds represented much of the increased density, whereas resident species exhibited no response. Density (individuals per hectare) of all species (n = 28) was not affected by the grazing treatment. Average total density was 40% greater in tebuthiuron-treated plots than in untreated plots. There was no overall tebuthiuron treatment or grazing effect on species richness. Diversity was lower on ungrazed, untreated plots than other treatment combinations in February and March. Increased density on tebuthiuron-treated plots was present in both wet and dry years. This finding differed from Martin (1990), who found no difference in relative abundance of all species between tebuthiuron-treated (using 0.5 kg/ha) and untreated shinnery oak communities in southeastern New Mexico.

Avian species richness, evenness, and diversity were only minimally affected by the tebuthiuron and grazing treatments (Smythe and Haukos 2010). Grasslands generally have low densities of birds, but estimated densities of this study were considerably lower than those reported in several other grassland studies (e.g., Cody, 1985; DeJong, 2001; Giezentanner, 1970; Igl & Ballard, 1999; Wiens, 1973). Current low densities in sand shinnery communities might indicate an ecological sink or reduced habitat carrying capacity. The moderate grazing regime (Holechek et al., 2001) of this study had little effect on grassland bird populations in this region, It is also important to emphasize that grazing was deferred on the study plots for 3 years after the tebuthiuron treatment, longer that the 1- to 2-year deferrals recommended by most current management guidelines (Peterson & Boyd, 1998). These results indicate that short-duration grazing regimes, based on the correct stocking rate and knowledge of available forage, are not detrimental to grassland bird populations in treated or untreated sand shinnery oak habitats.

Density of songbird nests (nests/10 ha) for all species was similar among treatments (Smythe & Haukos, 2009). The majority of nests of Cassin's sparrow (76%) and meadowlarks (90%) were in little bluestem (Smythe, 2006). Daily rate of nest survival across treatments did not differ between incubation and nestling period for any species or within any treatment. During incubation, daily survival of nests differed between tebuthiuron-treated and untreated plots as survival was 6.3% higher in untreated plots than in treated plots. However, during the nestling period the opposite trend was apparent as daily survival of nests was 17.3% greater in tebuthiuron-treated plots than in untreated plots.

In Smythe & Haukos (2009; 2010), the moderate grazing regime did not significantly impact vertical density of vegetation. Grassland birds selected nest sites based on overhead cover, presumably as a defense against avian predators; however, average overhead cover did not differ among treatments. Likewise, greater vertical cover in tebuthiuron-treated plots did not always result in higher daily rates of survival of nests. This may indicate that grasses and shrubs are needed during different periods of brood rearing and, thus, both are required in a restored shinnery oak community. Our results indicate that carefully managed application of tebuthiuron and grazing in shinnery oak communities do not adversely impact density or success of nests of grassland birds; however, current high rates of depredation and low rates of nest success overall do not bode well for grassland birds in this community.

Tebuthiuron treatment at low rates with appropriate grazing management may create a consistent grass/shrub mix normally restricted to years of above-average precipitation. This could increase densities of some migratory grassland bird species such as Cassin's Sparrow and does not appear to harm resident species. A carefully managed, moderate grazing regime also does not appear to negatively impact grassland bird density; however, grazing must be managed to maintain restoration efforts, and continued monitoring is needed to determine the long-term effects of restoration.

2. Conclusion

In areas where shinnery oak has become essentially a monoculture, tebuthiuron can be used to create vegetation heterogeneity that may benefit migratory grassland birds (Smythe & Haukos, 2009). A restored sand shinnery community then has a codominant mix of grass and shrubs. However, tebuthiuron treatment is not an excuse for continuing poor land management practices such as overgrazing. The goal of tebuthiuron use should be to increase the grass component within the shinnery oak community, not to eliminate shinnery oak entirely. It is important to realize that higher application rates desired by landowners to meet livestock goals may not be beneficial to grassland birds. Creating homogeneous grassland from homogeneous shrubland is likely not the best approach for grassland birds, and beneficial habitat for grassland birds could be created at lower application rates than occurred in this study. Tebuthiuron should not be applied to shinnery oak communities in poor condition, because on sandy soils, it can be difficult to obtain a high canopy cover of plants after sand shinnery oak is removed (Pettit, 1979). Adequate perennial bunchgrasses ($\geq$ 8 plants/m² suggested) must preexist in areas to be treated (Doerr, 1980; Jones, 1982).

Ethridge et al. (1987a) found that an application rate of 0.56 kg/ha was the maximum application that remained profitable. Doerr (1980) recommended a rate of 0.4 kg/ha to produce high densities and canopy coverage of forbs as well as grass seed as food for lesser prairie-chickens.

Grazing should be deferred for at least 1 year after tebuthiuron treatment to allow for adequate recovery; in this study, grazing was deferred for 3 years (Smythe & Haukos, 2009). Grazing too soon after application can result in serious erosion (Pettit, 1979). Grazing should be deferred for longer periods under drought conditions, and stocking rates should be calculated based on available forage and frequently reevaluated to maintain the benefit of tebuthiuron treatment. Grazing should be performed in a manner that mimics historic heterogeneous vegetation mosaics, where certain areas are grazed more intensively than others (Vickery et al., 1995). This will support a variety of grassland species that prefer different vegetation heights and densities. Short-duration grazing regimes may provide greater control over livestock effects on the landscape and accordingly make a heterogeneous vegetation mosaic easier to maintain.

Rainfall and the resulting vegetative conditions should always factor into any land management practice (Smythe & Haukos, 2009). The variability of weather in the Great Plains produces highly different conditions from year to year, and any management plan should be reevaluated each year to ensure that it is adequate for the conditions present. Knowledge of available forage is essential to calculating the correct stocking rate. It is critical to manage tebuthiuron-treated shinnery oak communities carefully to maintain the benefits of restoration. Carefully managing shinnery oak communities for bird populations can also provide benefits to humans, but the converse is not necessarily true.

3. References

Allison, C. (1994). Symposium on poisonous and noxious range plants: other poisonous plants. *Proceedings, Western Section, American Society of Animal Science* 45: 115-117.

Axelrod, D. (1985). Rise of the grassland biome, central North America. *Botanical Review* 51:163-201.

Bailey, J., & Painter, C. (1994). What good is this lizard? *New Mexico Wildlife* 39(4): 22-23.

Beckett, P. (1976). Mescalero Sands archaeological resource inventory. Second season. Report to the Bureau of Land Management, Roswell, New Mexico, Project YA-510-PH6-94. Report No. 45. Las Cruces: New Mexico State University.

Bednarz, J. (1987). The Los Medaños cooperative raptor research and management program; 1986 annual report. U. S. Department of Energy contract 59-WRK-90469-SD. Albuquerque: University of New Mexico, Department of Biology.

Bell, L., Fuhlendorf, S., Patten, M., Wolfe, D., & Sherrod, S. (2010). Lesser prairie-chicken hen and brood habitat use on sand shinnery oak. *Rangeland Ecology and Management* 63:478-486.

Biondini, M., Pettit, R., & Jones, V. (1986). Nutritive value of forages on sandy soils as affected by tebuthiuron. *Journal of Range Management* 39: 396-399.

Blair, W. (1950). The biotic provinces of Texas. *Texas Journal of Science* 2:93–117.

Bolen, E., Smith, L., & Schramm, H., Jr. (1989). Playa lakes: prairie wetlands of the Southern High Plains. Bioscience 39:615-623.

Bóo, R. (1974). Root carbohydrates in sand shinnery oak (*Quercus havardii* Rydb.) Master of Science thesis, Texas Tech University, Lubbock.

Bóo, R., & Pettit, R. (1975). Carbohydrate reserves in roots of sand shin-oak in west Texas. *Journal of Range Management* 28: 469-472.

Brown, D. E. (1982). Plains and Great Basin grasslands. In: D. E. Brown, editor. Biotic communities of the American Southwest United States and Mexico. *Desert Plants* 4: 115-121.

Brune, G. (2002). Springs of Texas, second edition. Texas A&M Press, College Station.

Bryant, F., & Demarais, S. (1992). Habitat management guidelines for white-tailed deer in south and west Texas. Research Highlights 1991 Noxious Brush and Weed Control; Range, Wildlife, & Fisheries Management 22: 11-13. Lubbock: Texas Tech University, College of Agricultural Sciences and Natural Resources.

Bryant, F., & Morrison, B. (1985). Managing plains mule deer in Texas and eastern New Mexico. Management Notes 7. Lubbock: Texas Tech Univ., College of Agricultural Sciences and Natural Resources.

Cody, M. (1985). Habitat selection in open-country birds. Pages 191-226 in M. L. Cody, editor, *Habitat selection in birds*. Academic Press, New York.

Colbert, R. (1986). The effect of the shrub component on small mammal populations in a sand shinnery oak ecosystem. M.S. Thesis, Texas Tech University, Lubbock.

Conner, N., Hyde, H., & Stoner, H. (1974). Soil survey of Andrews County, Texas. Washington, DC: U. S. Department of Agriculture, Soil Conservation Service.

Copelin, F. (1963). The lesser prairie chicken in Oklahoma. Technical Bulletin 6. Oklahoma City: Oklahoma Wildlife Conservation Department.

Crawford, J. (1974). The effects of land use on lesser prairie chicken population in west Texas. Ph.D. Dissertation, Texas Tech University, Lubbock.

Crawford, J. (1980). Status, problems, and research needs of the lesser prairie chicken. pp. 1-7. In: P. A. Vohs and F. L. Knopf, editors, *Proceedings of the Prairie Grouse Symposium*. Stillwater: Oklahoma State University.

Crawford, J., & Bolen, E. (1976). Fall diet of lesser prairie chickens in west Texas. *Condor* 78: 142-144.

Dayton, W. (1931). Important western browse plants. Miscellaneous Publication 101. Washington, DC: U. S. Department of Agriculture.

Davis, C., Riley, T., Smith, R., Suminski, H., & M.D. Wisdom. (1979). Habitat evaluation of lesser prairie chickens in eastern Chaves County, New Mexico, New Mexico. New Mexico State University, Agricultural Experiment Station, Las Cruces.

DeJong, J. (2001). Landscape fragmentation and grassland patch size effects on non-game grasslands birds in xeric mixed-grass prairies of western South Dakota. Master of Science Thesis, South Dakota State University, Brookings.

Deering, D., & Pettit, R. (1972). Sand shinnery oak acreage survey. Research Highlights Noxious Brush and Weed Control; Range, Wildlife, & Fisheries Management 2: 14. Lubbock: Texas Tech University, College of Agricultural Sciences and Natural Resources.

Degenhardt, W., & Jones, K. (1972). A new sagebrush lizard, *Sceloporus graciosus*, from New Mexico and Texas. *Herpetologica* 28:212-217.

Degenhardt, W., Painter, C. & Price, A. (1996). *Amphibians and reptiles of New Mexico*. Albuquerque, University of New Mexico Press.

Dhillion, S., McGinley, M., Friese, C., & Zak, J. (1994). Construction of sand shinnery oak communities of the Llano Estacado: animal disturbances, plant community structure, and restoration. *Restoration Ecology* 2:51-60.

Dickerson, R. (1985). Short duration versus continuous grazing on sand shinnery oak range. Master of Science Thesis, Texas Tech University, Lubbock, Texas, USA.

Doerr, T. (1980). Effects of tebuthiuron on lesser prairie chicken habitat and food supplies. M.S. Thesis, Texas Tech University, Lubbock.

Doerr, T., & Guthery, F. (1980). Effects of shinnery oak control on lesser prairie chicken habitat. pp. 59-63. In: P. A. Vohs and F. L. Knopf, editors, *Proceedings of the Prairie Grouse Symposium*. Stillwater: Oklahoma State University. 89 pp.

Doerr, T., & Guthery, F. (1983). Effects of tebuthiuron on lesser prairie chicken habitat and foods. *Journal of Wildlife Management* 47:1138-1142.

Dollahite, J. (1961). Shin-oak (*Quercus havardi*) poisoning in cattle. *Southwestern Veterinarian* 14: 198-201.

Donaldson, D. (1966). Brush control and the welfare of lesser prairie chickens in western Oklahoma. *Oklahoma Academy of Science Proceedings for 1965* 46: 221-228.

Donaldson, D. (1969). Effect on lesser prairie chickens of brush control in western Oklahoma. Ph.D. thesis, Oklahoma State University, Stillwater.

Duck, L. & Fletcher, J. (1944). A survey of the game and fur-bearing animals of Oklahoma. State Bulletin 3. Oklahoma City: Oklahoma Game and Fish Department.

Emmerich, W. (1985). Tebuthiuron - environmental concerns. *Rangelands* 7: 14-16.

Ethridge, D., Pettit, R., Neal, T., & Jones, V. (1987)a. Economic returns from treating sand shinnery oak with tebuthiuron in west Texas. *Journal of Range Management* 40: 346-348.

Ethridge, D., Pettit, R., Sudderth, R., & Stoecker, A. (1987)b. Optimal economic timing of range improvement alternatives: southern High Plains. *Journal of Range Management* 40: 555-559.

Fischer, N. (1985). Wildlife. pp. 57-72. In: N. T. Fischer, editor. *Ecological Monitoring Program at the Waste Isolation Pilot Plant*. Second Semiannual Report Covering Data Collected January to June 1985. DOE/WIPP-85-002. Carlsbad, NM: U. S. Department of Energy. 128 pp.

Frary, L. (1957). Evaluation of prairie chicken ranges. Completion Reports Project W-77-R-3. Santa Fe: New Mexico Department of Game and Fish.

Galbraith, J. (1983). Plant and soil water relationships following sand shin-oak control. Master of Science thesis, Texas Tech University, Lubbock.

Garrison, G., Bjugstad, A., Duncan, D., Lewis, M., & Smith, D. (1977). Vegetation and environmental features of forest and range ecosystems. Agriculture Handbook 475. Washington, DC: US Department of Agriculture, Forest Service.

Giezentanner, J. (1970). Avian distribution and population fluctuations on the shortgrass prairie of north central Colorado. Master of Science Thesis, Colorado State University, Fort Collins.

Gile, L. (1979). Holocene soils in eolian sediments of Bailey County, Texas. *Soil Science Society of America Journal* 43:994-1003.

Gorum, L., Snell, H., Pierce, L., & McBride, T. (1995). Results from the fourth year (1994) research on the effect of shinnery oak removal on the dunes sagebrush lizard, *Sceloporus arenicolus*, in New Mexico. Report submitted to the New Mexico

Department of Game and Fish; contract #80-516.6-01. Albuquerque: University of New Mexico, Museum of Southwestern Biology.

Graul, W. (1980). Grassland management practices and bird communities. Pages 38-47 *in* DeGraff, R.M., and T.G. Tilgham, compilers. *Management of western forests and grasslands for nongame birds*. General Technical Report INT-86. United States Forest Service, Intermountain Forest and Range Experiment Station, Rocky Mountain Forest and Range Experiment Station, and Intermountain Region, Ogden, Utah.

Gray, G., Hampy, D., Simpson, C., Scott, G., & Pence, D. (1978). Autumn rumen contents of sympatric Barbary sheep and mule deer in the Texas Panhandle. Research Highlights 1977 Noxious Brush and Weed Control; Range, Wildlife, & Fisheries Management 8: 34. Lubbock: Texas Tech University, College of Agricultural Sciences and Natural Resources.

Greer, H., McIlvain, E., & Armstrong, C. (1968). Controlling shinnery oak in western Oklahoma. OSU Extension Facts No. 2765. Stillwater: Oklahoma State University.

Gross, F.. & Dick-Peddie, W. (1979). A map of primeval vegetation in New Mexico. *Southwestern Naturalist* 24: 115-122.

Hafsten, U. (1961). Pleistocene development of vegetation and climate in the southern High Plains as evidenced by pollen analysis. pp. 59-91. In: F. Wendorf, editor, *Paleoecology of the Llano Estacado*. Santa Fe, NM: Museum of New Mexico Press. Fort Burgwin Research Center Publication 1.

Haukos, D., & Smith, L. (1989). Lesser prairie-chicken nest site selection and vegetation characteristics in tebuthriuon treated and untreated sand shinnery oak in Texas. *Great Basin Naturalist* 49:624-626.

Haukos, D., & McDaniel, P. (2011). Results of long-term monitoring of lesser prairie-chicken habitat on the Milnesand Prairie Preserve, The Nature Conservancy of New Mexico. The Nature Conservancy, Santa Fe, New Mexico.

Herbel, C. H. 1979. Utilization of grass- and shrublands of the southwestern United States. pp. 161-203. In: B. H. Walker, Management of Semi-arid Ecosystems. New York: Elsevier Scientific Publishing Company.

Herndon, E. (1981). Shredding detrimental to herbicide activity on sand shinnery oak. Research Highlights 1980 Noxious Brush and Weed 34 USDA Forest Service Gen. Tech. Rep. RMRS–GTR–16. 1998 Control; Range, Wildlife, & Fisheries Management 11: 62. Lubbock: Texas Tech University, College of Agricultural Sciences and Natural Resources.

Hodson, M., Calhoun, T., Chastain, C., Hacker, L., Henderson, W., & Seagraves, C. R. (1980). Soil survey of Chaves County, New Mexico, southern part. Washington, DC: USDA Soil Conservation Service. 148 pp. + maps.

Holechek, J., Pieper, R., & Herbel, C. (2001). Range management: principles and practices. Prentice Hall, Upper Saddle River, New Jersey.

Holland, M. (1994). Disturbance, environmental heterogeneity, and plant community structure in a sand shinnery oak community. Master of Science thesis, Texas Tech University, Lubbock.

Holliday, V. (1989). Middle Holocene drought on the Southern High Plains. *Quaternary Research* 31:74-82.

Holliday, V. (1990). Soils and landscape evolution of eolian plains: the Southern High Plains of Texas and New Mexico. *Geomorphology* 3:489-515.

Holliday, V. (2001). Stratigraphy and geochronology of upper Quaternary eolian sand on the Southern High Plains of Texas and New Mexico, United States. *Geological Society of America Bulletin* 113:88-108.

Jacoby, P., Slosser, J., & Meadows, C. (1983). Vegetational responses following control of sand shinnery oak with tebuthiuron. *Journal of Range Management* 36: 510-512.

Johnson, K., Smith, B., Sadoti, G., Neville, T. & Neville, P. (2004). Habitat use and nest site selection by lesser prairie-chickens in southeastern New Mexico. *Southwestern Naturalist* 49:334-343.

Johnson, P., & Ethridge, D. (1996). The value of brush control and related research at Texas Tech University to the state of Texas. Research Highlights 1995 Noxious Brush and Weed Control; Range, Wildlife, & Fisheries Management 26: 12. Lubbock: Texas Tech University, College of Agricultural Sciences and Natural Resources.

Jones, V., & Pettit, R. (1980). Soil water in Graslan® and untreated oak plots. Research Highlights 1979 Noxious Brush and Weed Control; Range, Wildlife, & Fisheries Management 10: 31. Lubbock: Texas Tech University, College of Agricultural Sciences and Natural Resources.

Jones, V. (1982). Effects of tebuthiuron on a sand shinnery oak community. Ph.D. dissertation, Texas Tech University, Lubbock.

Jones, V. & Pettit, R. (1982). Graslan® provides long term sand shinoak control. Research Highlights 1981 Noxious Brush and Weed Control; Range, Wildlife, & Fisheries Management 12: 21. Lubbock: Texas Tech University, College of Agricultural Sciences and Natural Resources.

Jones, V., & Pettit, R. (1984). Low rates of tebuthiuron for control of sand shinnery oak. *Journal of Range Management* 37: 488-490.

Krysl, L., Simpson,C., & Gray, G. (1980). Dietary overlap of sympatric Barbary sheep and mule deer in Palo Duro Canyon, Texas. pp. 97- 103. In: Simpson, C. D., editor, Proceedings of the Symposium on ecology and management of Barbary sheep, November 19-21, 1979. Lubbock: Texas Tech University, Department of Range & Wildlife Management.

Kuchler, A. (1970). Potential natural vegetation, in *The National Atlas of the United States of America*: U.S. Geological Survey, p. 89–92.

Lenfesty, C. (1983). Soil survey of Chaves County, New Mexico, Northern Part. Washington, DC: U. S. Department of Agriculture, Soil Conservation Service.

Ligon, J. (1927). Wild life of New Mexico. Its conservation and management.Santa Fe, NM: State Game Commission.

Lotspeich, F., & Everhart, M. (1962). Climate and vegetation as soil forming factors on the Llano Estacado. *Journal of Range Management* 15: 134-141.

MacGinitie, H. (1962). The Kilgore flora: a late Miocene flora from northern Nebraska. *University of California Publication of Geological Science* 35:67-158.

McIlvain, E. (1954). Interim report on shinnery oak control studies in the southern Great Plains. pp. 95-96. In: Proceedings, *Eleventh Annual Meeting, North Central Weed Control Conference*, December 6-9, 1954, Fargo, ND.

McIlvain, E., & Armstrong, C. (1959). Shinnery oak control produces more grass. Preprint for *Proceedings, Southern Weed Conference*, January, 1959, Shreveport, LA.

Marcy, R. (1850). In compliance with a resolution of the House of the 6th February, a report and map of Lt. Simpson, of the route from Fort Smith to Santa Fe; also, a report on the same subject from Captain R.B. Marcy, 5th Infantry. Washington, D.C.: 31st Congress, 1st Session, House of Representatives Executive Document 7 No. 45.

Martin, B. (1990). Avian and vegetation research in the shinnery oak ecosystem of southeastern New Mexico. M.S. thesis. New Mexico State University, Las Cruces.

Mayes, S. (1994). Clonal population structure of *Quercus havardii* (sand shinnery oak). Master of Science thesis, Texas Tech University,Lubbock.

Moldenhauer, W., Coover, J., & Everhart, M. (1958). Control of wind erosion in the sandy lands of the southern High Plains of Texas and New Mexico. ARS 41-20. Washington, DC: US Department of Agriculture, Agricultural Research Service.

Morris, J. (1997). El Llano Estacado, Exploration and Imagination on the High Plains of Texas and New Mexico, 1536-1860. Texas State Historical Association, Austin.

Muhs, D., & Holliday, V. (1995). Evidence of active dune sand on the Great Plains in the 19th century from accounts of early explorers. *Quaternary Research* 43:198-208.

Muhs, D., & Holliday, V. (2001). Origin of late Quaternary dune fields on the Southern High Plains of Texas and New Mexico. *Geological Society of America Bulletin* 113:75-87.

Muller, C. (1951). The significance of vegetative reproduction in *Quercus*. *Madroño* 11: 129-137.

Newman, A. (1964). Soil survey of Cochran County, Texas. Washington, DC: U. S. Department of Agriculture, Soil Conservation Service. 80 pp. + maps.

(NMPIF) New Mexico Partners in Flight. (2007). Bird Conservation Plan version 2.1 Albuquerque, New Mexico, USA (available from http:// www.nmpartnersinflight.org/bcp.html) accessed 9 December 2007.

Olawsky, C. (1987). Effects of shinnery oak control with tebuthiuron on lesser prairie-chicken populations. Master of Science thesis, Texas Tech University, Lubbock. 8

Olawsky, C., Smith, L., & Pettit, R. (1988). Effects of shinnery oak control on early summer diet and condition of lesser prairie-chickens. Research Highlights 1987 Noxious Brush and Weed Control; Range, Wildlife, & Fisheries Management 18: 29. Lubbock: Texas Tech University, College of Agricultural Sciences and Natural Resources.

Olawsky, C., & Smith, L. (1991). Lesser prairie-chicken densities on tebuthiuron-treated and untreated sand shinnery oak rangelands. *Journal of Range Management* 44: 364-368.

Parks, H. (1937). Valuable plants native to Texas. Bulletin 551. College Station: Texas Agricultural Experiment Station.

Patten, M., Wolfe, D., Shochat, E., & Sherrod, S. (2005). Effects of microhabitat and microclimate selection on adult survivorship of the lesser prairie chicken. *Journal of Wildlife Management* 69:1270-1278.

Peterson, R., & Boyd, C. (1998). Ecology and management of sand shinnery communities: a literature review. General Technical Report RMRS-GTR-16. USDA Forest Service, Fort Collins, Colorado.

Pettit, R. (1975). Histological studies of sand shin-oak leaves. Research Highlights 1974 Noxious Brush and Weed Control; Range, Wildlife, & Fisheries Management 5: 29. Lubbock: Texas Tech University, College of Agricultural Sciences and Natural Resources.

Pettit, R. (1976). Further evaluation of picloram on sand shin-oak communities. Research Highlights 1975 Noxious Brush and Weed Control; Range, Wildlife, & Fisheries

Management 6: 40. Lubbock: Texas Tech University, College of Agricultural Sciences and Natural Resources.

Pettit, R. (1977). The ecology and control of sand shin-oak. Ranch Management Conference, September 23, 1977 [Proceedings] 15: 6-11.Texas Tech University, Lubbock.

Pettit, R. (1978). Soil-vegetation relationships on dune sands. Research Highlights 1977 Noxious Brush and Weed Control; Range, Wildlife, & Fisheries Management 8: 10. Lubbock: Texas Tech University, College of Agricultural Sciences and Natural Resources.

Pettit, R. (1979). Effects of picloram and tebuthiuron pellets on sand shinnery oak communities. *Journal of Range Management* 32:196-200.

Pettit, R. (1986). Sand shinnery oak: control and management. Management Note 8. Lubbock: Texas Tech University, Range and Wildlife Management. 5 pp.

Pettit, R. (1994). Sand shinnery oak. p. 106. In: T. N. Shiflet, editor, *Rangeland Cover Types of the United States*. SRM 730. Denver: Society for Range Management.

Plumb, G. (1984). Grazing management following sand shin-oak control. Master of Science thesis, Texas Tech University, Lubbock.

Plumb, G., & Pettit, R. (1983). Grazing study on sand shinnery oak range. Research Highlights 1982 Noxious Brush and Weed Control; Range, Wildlife, & Fisheries Management 13: 49-51. Lubbock: Texas Tech University, College of Agricultural Sciences and Natural Resources.

Raught, R. (1967). White-tailed deer. pp. 52-60. In: *New Mexico Wildlife Management*. Santa Fe: New Mexico Department of Game and Fish.

Reeves, C. Jr. (1965). Pleistocene climate of the Llano Estacado. *Journal of Geology* 73: 181-189.

Riley, T., Davis, C., Ortiz, M., & Wisdom, M. (1992). Vegetative characteristics of successful and unsuccessful nests of lesser prairie chickens. *Journal of Wildlife Management* 56:383–387.

Riley, T., Davis, C., & Smith, R. (1993). Autumn-winter habitat use of lesser prairie-chickens (*Tympanuchus pallidicinctus*, Tetraonidae). *Great Basin Naturalist* 53: 409-411.

Roebuck, C. (1982). Comparative food habits and range use of pronghorn and cattle in the Texas Panhandle. Master of Science thesis, Texas Tech University, Lubbock.

Roebuck, C., & Simpson, C. (1982). Food habits of pronghorn antelope and cattle in the Texas Panhandle. Research Highlights 1981 Noxious Brush and Weed Control; Range, Wildlife, & Fisheries Management 12: 76. Lubbock: Texas Tech University, College of Agricultural Sciences and Natural Resources.

Rowell, C. (1967). Vascular plants of the Texas Panhandle and South Plains. Ph.D. thesis, Oklahoma State University, Stillwater.

Scifres, C., & Mutz, J. (1978). Herbaceous vegetation changes following applications of tebuthiuron for brush control. *Journal of Range Management* 31:375-378.

Scifres, C., Stuth, J., & Bovey, R. (1981). Control of oaks (*Quercus* spp.) and associated woody species on rangeland with tebuthiuron. *Weed Science* 29: 270-275.

Sears, W., Britton, C., Wester, D., & Pettit, R. (1986). Herbicide conversion of a sand shinnery oak (*Quercus havardii*) community: effects on biomass. *Journal of Range Management* 39: 399-403.

Sell, D. (1979). Spring and summer movements and habitat use by lesser prairie chicken females in Yoakum County, Texas. Master of Science thesis, Texas Tech University, Lubbock.

Sikes, D., &Pettit, R. (1980). Soil temperature, oxygen, and water level effects on sand shinnery oak. *Soil Science* 130: 344-349.

Slosser, J., Jacoby, P., & Price, J. (1985). Management of sand shinnery oak for control of boll weevil (Coleoptera: Curculionidae) in the Texas Rolling Plains. *Journal of Economic Entomology* 78: 383-389.

Smith, C. (1985). To save a dune. *Southwest Heritage* 14(1): 5-8, 12, 19.

Smith, L. (2003). *Playas of the Great Plains*. University of Texas Press, Austin.

Smythe, L. (2006). Response of nesting grassland birds to sand shinnery oak communities treated with tebuthiuron and grazing in eastern New Mexico. M.S. thesis, Texas Tech University, Lubbock.

Smythe, L., & Haukos, D. (2009). Nesting success of grassland birds in shinnery oak communities treated with tebuthiuron and grazing in eastern New Mexico. *Southwestern Naturalist* 54:136-145.

Smythe, L., & Haukos, D. (2010). Response of grassland birds in sand shinnery oak communities restored using tebuthiuron and grazing in eastern New Mexico. *Restoration Ecology* 18:215-223.

Snell, H., Gorum, L., Pierce, L., & Ward, K. (1997). Results from the fifth year (1995) research of the effect of shinnery oak removal on populations of sand dune lizards, Sceloporus arenicolus, in New Mexico. Report submitted to the New Mexico Department of Game and Fish, contract #80-516.6-01. Albuquerque. University of New Mexico, Museum of Southwestern Biology.

Stebbins, G. (1981). Coevolution of grasses and herbivores. *Annual Publication of Missouri Botanical Gardens* 68:75-86.

Sullivan, J. (1980). Differentiation of sand shinnery oak communities in west Texas. Master of Science thesis, Texas Tech University, Lubbock.

Sullivan, J., & Pettit, R. D. (1977). Mapping of sand shin oak. Research Highlights 1976 Noxious Brush and Weed Control; Range, Wildlife, & Fisheries Management 7: 31. Lubbock: Texas Tech University, College of Agricultural Sciences and Natural Resources.

Taylor, M., & Guthery, F. (1980). *Status, ecology, and management of the lesser prairie chicken*. General Technical Report RM-77. Fort Collins: US Department of Agriculture, Forest Service. 15 pp.

Thurmond, M., Pettit, R., & Fryrear, D. (1986). Evaluation of wind erosion on sandy rangelands. Research Highlights 1985 Noxious Brush and Weed Control; Range, Wildlife, & Fisheries Management 16: 27. Lubbock: Texas Tech University, College of Agricultural Sciences and Natural Resources.

U. S. Department of the Interior Bureau of Land Management. (1977). *Vegetative description and analysis from literature reviews for shinnery oak (Quercus havardii) vegetative community in the Roswell BLM District*. Roswell, NM.

Wester, D. (2007). The Southern High Plains: a history of vegetation from 1540 to present. Pages 24-47 In: Sosebee, R.E.; Wester, D.B.; Britton, C.M.; McArthur, E.D.; Kitchen, S.G., comp. 2007. *Proceedings: Shrubland dynamics — fire and water*; 2004 August 10-12; Lubbock, TX. Proceedings RMRS-P-47. Fort Collins, CO: U.S. Department of Agriculture, Forest Service, Rocky Mountain Research Station.

Wiens, J. (1973). Pattern and process in grassland bird communities. *Ecological Monographs* 43:237–270.

Wiedeman, V. (1960). Preliminary ecological study of the shinnery oak area of western Oklahoma. Master of Science thesis, University of Oklahoma, Norman.

Willig, M., Colbert, R., Pettit, R., & Stevens, R. (1993). Response of small mammals to conversion of a sand shinnery oak woodland into a mixed mid-grass prairie. *Texas Journal of Science* 45: 29-43.

Wolfe, H., editor. (1978). An environmental baseline study on the Los Medaños Waste Isolation Pilot Plant (WIPP) project area of New Mexico: a progress report. Sandia Laboratory Studies SAND 77-7017. Albuquerque: U.S. Department of Energy. 112 pp.

Wright, H., & Bailey, A. (1982). Fire Ecology, United States and Southern Canada. Wiley-Interscience, NY.

Zhang, Q. (1996). Fungal community structure and microbial biomass in a semi-arid environment: roles in root decomposition, root growth, and soil nitrogen dynamics. Ph.D. thesis, Texas Tech University.

Zobeck, T., Fryrear, D., & Pettit, R. (1989). Management effects on wind-eroded sediment and plant nutrients. *Journal of Soil and Water Conservation* 44: 160-163.

8

Herbicides and the Risk of Neurodegenerative Disease

Krithika Muthukumaran, Alyson J. Laframboise
and Siyaram Pandey
University of Windsor
Canada

1. Introduction

In the quest for increased agricultural productivity, longer shelf life of produce, weed-free lawns and sanitized facilities, we have developed a plethora of pesticides. Pesticides are a broad range of substances commonly used to control insects, weeds, and fungi (plant diseases). They are classified by target organism or mode of use as insecticides, herbicides, fungicides, or fumigants. Development, manufacturing and large scale application of these products have become main stream practices in industry, agriculture and domestic sectors. More than 18,000 products are registered for use in the United States, and > 1 billion pounds of pesticides are applied annually as of 2007 (EPA, 2011). Any chemical designed to kill weeds or insects certainly has potential to harm humans. Short term exposures to low dosages of these chemicals are generally non-toxic. Detoxification systems in the human body are capable of modifying and clearing these molecules from the system efficiently. However, long term exposure to these toxic chemicals combined with poor drug catabolizing cytochrome p450 activity in some individual could lead to its accumulation in the system and toxicity leading to increased risk of certain diseases. Many of the herbicides are toxic to mitochondria and cause oxidative stress. Neuronal cells, being critically dependent on mitochondrial function and sensitive to oxidative stress, often fall victim to these herbicides. Slow and progressive loss of neurons leads to neurodegenerative diseases. In this review we describe toxic effects, mechanism of action and results of animal studies of selected herbicides implicated in neurodegenerative diseases. We also discuss the need for stringent testing of these kinds of substances for neurotoxicity and recent advances in neuroprotective therapies as outlined in the overview before.

2. Specific herbicides with neurodegenerative effects

2.1 Paraquat

2.1.1 Background and history of use

Paraquat is the trade name of N,N′-dimethyl-4,4′-bipyridinium chloride, the dichloride salt of the radical 1,1′-dimethyl-4,4′-dipyridilium. It is one of the most commonly used and powerful herbicides and was synthesised in 1932 at the Rockefeller Institute by Michaelis

(Michaelis & Hill, 1933). In the early years it was called methyl viologen because it readily reduced to a stable blue or violet free radical and this property was exploited by biochemists who used the compound as an oxidation-reduction indicator. In 1955, the herbicidal properties of bipyridils were researched by Imperial Chemical Industries Ltd and the herbicidal nature of paraquat was established (Smith & Heath, 1976). In the year 1962, paraquat was introduced to the market as a herbicide by the Plant Protection Division of Imperial Chemical Industries Ltd, which is now known as Syngenta (www2.syngenta.com/en/about_syngenta/companyhistory.html). It is the third best-selling pesticide in the world (www2.syngenta.com), and is especially popular in the developing countries (Wesseling et al., 2001).

Paraquat is extensively used all over the world because it is non-selective, fast acting, and small quantities are sufficient to efficiently kill weeds, thereby reducing competition for light, water and nutrients. Through use of paraquat, fields can be quickly prepared for farming, and rain does not affect the action of the herbicide. Paraquat is used on over 100 crops in more than 120 countries across the world (Wessling et al., 2001).

2.1.2 Toxic effects and mechanism of action

Paraquat is a non-selective herbicide that requires oxygen and light energy for its action. In green plants, the light energy captured by chlorophyll is transferred as electrons through photosystem I. In the presence of the reduced form of paraquat, the electrons from the Fe-S centres of photosystem I are diverted and react with paraquat (Conning et al., 1969) to generate superoxide anion O_2^-. This in turn generates hydroxyl radicals either directly or via the intermediary hydrogen peroxide. These highly reactive radicals cause deterioration of the cell membrane leading to cellular breakdown (Matile & Moor, 1968). Browning of leaves is seen in a few hours in the presence of strong light and complete desiccation is observed in a few days. Fewer cases of resistance have been observed in comparison with resistance to herbicides with other modes of action.

Paraquat is extremely poisonous and use of the herbicide requires a licence in most countries. Use of paraquat has been banned in Finland, Austria, Sweden and Norway because of its high toxicity and frequency of poisonings (Wesseling et al., 2001). Paraquat manufactured by Syngenta, the leading producer of this herbicide, is blue in colour, has a sharp odour and contains an agent that causes vomiting, which helps prevent accidental consumption of the herbicide. Ingestion of a high dose of paraquat causes lung congestion, difficulty in breathing and an increase in heart rate. Cases of lung scarring have been reported when marijuana contaminated with paraquat was consumed. When the eyes come in contact with paraquat, it can cause corneal damage and scarring, while contact with skin leads to burns, damage to fingernails and dermatitis. Paraquat has been shown to be a mutagen to mouse, human and microorganisms; however there was no mutation observed in the sperm of male mice (Hazardous Substances Databank, 1995). There have been no adverse effects on reproduction or birth defects reported so far.

Paraquat toxicity depends on the amount, route and duration of exposure as well as the person's health at the time of exposure. Inhalation of paraquat can lead to acute respiratory distress syndrome whereas ingestion can cause direct damage to the lining of mouth or intestines. When ingested, it is distributed all over the body and toxic changes occur primarily in the lungs, liver and kidneys (Wagner, 1981). In humans, the lethal dose for ingestion is 35 mg/Kg. Those suffering from respiratory infection are more susceptible to the toxic effects of paraquat. There is a high accumulation of paraquat in the lung tissue compared to other organs (Stevens & Sumner, 1991).

Paraquat is exceedingly toxic to aquatic species such as rainbow trout, bluegill and channel catfish. At high concentration, it is shown to inhibit photosynthesis in certain types of algae. It is not known to accumulate in tissue; however, it can bioaccumulate in weeds. Paraquat can be found in residual form bound to aquatic weeds and bottom mud (Weed Science Society of America, 1994).

The chronic effects of paraquat in humans are Parkinson's Disease (PD) and severe lung damage (Wagner, 1981). In order to understand the environmental factors that make one susceptible to PD, a study of 120 patients was carried out in Taiwan. The study included 240 controls that were selected based on age and sex. The interview comprised questions related to the history of exposure to environmental factors such as source of drinking water, probability of environmental or occupational exposure, and use of herbicides or pesticides or exposure to paraquat. The results showed that exposure to paraquat caused an increase in susceptibility to PD (Liou, 1997).

A study from California showed that consumption of well water contaminated with paraquat increased susceptibility to PD by 20-50%. Almost 91% of the study population was exposed to well water contaminated with paraquat (Gatto et al., 2009).

The predisposing factors for complex diseases such as PD are both genetic and non-genetic. Hence, it is necessary to study the joint effects of both these factors and in the case of genetic factors, the genes that are responsible for increasing the susceptibility to PD and the pathways these genes are involved in have to be identified. Out of the 1460 single nucleotide polymorphisms in the brain that have been identified in the axon guidance pathway, 183 (12.5%) have been shown to increase susceptibility to PD (Lesnick, 2007).

2.1.3 Development of animal models

In 1983, Langston reported irreversible Parkinsonism (Langston et al., 1983) in a group of young drug addicts who used synthetic heroin contaminated with 1-methyl-4 phenyl-1,2,3,6- tetrahydropyridine (MPTP). Since the contaminant caused symptoms of PD, monkey (Burns et al., 1983) and mouse (Heikkila et al., 1984) animal models of PD were developed by injecting sub lethal doses of MPTP. Animal models provide an opportunity to study the pathophysiology of PD and assist in development of successful treatment and prevention strategies. MPTP is metabolized into MPP+, which is structurally similar to paraquat. Since exposure to MPTP is minimal, but paraquat is a risk factor for PD, an animal model using paraquat is helpful for understanding the pathophysiology of PD.

Rat and mouse animal models have been developed to study the effect of exposure to paraquat, especially the neurotoxic effects. The toxic effect of paraquat on dopaminergic neurons is slow and this is reflected by the delayed and slow progression in the pathogenesis of PD. Both long and short term studies have been carried out on rats. Evans hooded rats showed loss of dopaminergic neurons in the substantia nigra region of the brain when intraperitoneal injections of paraquat were given weekly for three weeks. The neurons of the substantia nigra pars compacta region of the brain, which are autonomic pacemakers, produce high levels of reactive oxygen species and free radicals due to the accumulation of intracellular calcium. This results in DNA damage and high levels of neurodegeneration in this region. The paraquat-induced loss of neurons was shown by comparing tyrosine hydroxylase immunohistochemistry in the substantia nigra of paraquat injected rats with the substantia nigra of saline injected control rats. Loss of approximately 70% of the neurons was observed in the paraquat injected rats. Though not many differences can be seen visually, behavioural assessment using rotorod showed measurable changes between the two groups. Biochemical assays have shown that paraquat induces oxidative stress, indicating its neurotoxic effects. The reduced form of paraquat has been shown to cause an increase in reactive oxygen species (ROS) which is selective to brain cells (Castello et al., 2007). Subtle behavioural differences were also observed in the test and control rats (Somayajulu- Nitu et al., 2009). When a similar dose was given as intraperitoneal injections to Wistar rats for 4, 6, 8, 12, and 24 weeks and immunohistochemistry studies performed, a loss of approximately 37% in the dopaminergic neurons was observed (Ossowska et al., 2005).

Dopaminergic neuron loss in the substantia nigra and intra-neuronal deposition of α-synuclein containing aggregates were also observed when C57BLJ/6 mice were injected intraperitoneally with paraquat (10mg/Kg body weight) weekly for three consecutive weeks (Manning-Bog et al., 2001). Similar studies have also been carried out with transgenic mice over-expressing α-synuclein and similar results have been obtained (Fernagut et al., 2007). The animal models help support the epidemiological link between paraquat used for agricultural purposes and PD. The C57BL/6 mouse strain has also been used for toxicology studies. When paraquat solution was applied to these mice through their nares, they showed lung pathology similar to those suffering from paraquat toxicity (Tomita et al., 2007).

2.2 Rotenone

2.2.1 Background and history of use

Rotenone is an insecticide and piscicide extracted from tropical leguminose. It was first isolated by Emmanual Geoffrey from the plant *Robinia nicou* while he was travelling in French Guiana. The details of the compound, which he called nicouline, were published posthumously in his thesis in 1895 after he died due to parasitic infection. However, rotenone has been used for many years by Africans and Asians to intoxicate fish and kill caterpillars.

Rotenone is now used in commercial gardens and is found in animal care products. It is marketed as an organic pesticide and piscicide and is available as dust, powder and spray formulations. The World Health Organisation has assigned rotenone as a moderately hazardous chemical; in Canada and USA, it is used only as a piscicide. It is also found in formulation with other pesticides such as carbaryl and lindane. It is manufactured by many companies and is available in 300 formulated products. It is effective against a large number of insect pests such as apple maggot, European corn borer, Pea aphid, Japanese beetle, Ladybird beetles, and predatory mites.

2.2.2 Toxic effects and mechanism of action

Complex I, which is present in the inner mitochondrial membrane, is the first enzyme of the electron transport chain. This enzyme is encoded by the mitochondrial genome, unlike the other enzymes which are encoded by the nuclear genome (Hatefi, 1985). Rotenone inhibits complex I of the respiratory chain, resulting in a decrease in energy production.

The toxicity of rotenone depends on the nature of the plant extract and the species from which it is extracted. It is mildly toxic to hamsters or mice, whereas it is highly toxic to rats, especially the females. For humans the toxic dose is 300-500 mg/Kg for adults and 143 mg/Kg for children. Human fatality due to rotenone has been reported very rarely because it causes immediate vomiting upon consumption. However, when the dust is inhaled it could lead to an increase in the rate of respiration followed by depression and convulsions (Uversky, 2004).

There have been no reports of endocrine disruption or teratogenic effects in humans. Studies on the carcinogenic effects of rotenone are still inconclusive. Studies on animal models of rotenone however show that it can produce symptoms of PD.

2.2.3 Development of animal models

Rotenone is involved in the pathogenesis of PD. Animal models of PD by exposure to various doses of rotenone for different durations have been developed to study the relationship between complex I and PD and also for preclinical testing of neuroprotective strategies. A number of genetic mouse models of PD are being developed; however, they fail to show the exact pathophysiology of PD. Hence, exposing the genetic mice to rotenone could help understand the link between genetic and environmental factors in the development of PD as well.

Sprague Dawley and Lewis rats were injected with varying doses of rotenone and it was shown that Lewis rats give more consistent results. A high dose of rotenone caused cardiovascular failure and non-specific brain lesions (Ferrante et al., 1997). However, a dose of 2-3 mg/Kg given for four weeks directly into the veins or subcutaneously resulted in PD-like symptoms. Rotenone was shown to act by inhibiting complex I, resulting in a decrease in the dopaminergic neurons in the substantia nigra region of the brain and also dopaminergic lesions (Hoglinger et al., 2005). The behavioural changes include unsteady movement and hunched posture. Severe rigidity was observed in some, whereas resting tremors were observed in others. Rotenone also caused loss of non- dopaminergic neurons in the basal ganglia and brainstem. The effect of rotenone was on neurons and not on oligodendrocytes, astrocytes or microglial cells. The drawback of the rotenone animal model is the variability in the effects of the same dose of rotenone, which clearly indicates the differences in the sensitivity of different strains of rat to the pesticide.

A group in Japan has shown that C57BL/6 mice, when orally fed with 30 mg/Kg of rotenone for 56 days, developed motor deficits, neurodegeneration of dopaminergic neurons and increases in the cytoplasmic accumulation of α- synuclein in the remaining dopaminergic neurons (Inden et al., 2011). Defects in complex I lead to a decrease in ATP synthesis which in turn induces mitochondrial depolarization and calcium deregulation. These factors cause an increase in the production of reactive oxygen species and hence oxidative stress (Sohal & Weindruch, 1996). This, along with various genetic factors, results in the progression of neurodegenerative diseases. It has been shown that there is a 15-30% decrease in the activity of complex I in sporadic cases of PD (Betarbet et al., 2000). Rotenone, being lipophilic, crosses the cellular membranes and the blood brain barrier, becomes accumulated in the sub-cellular organelles such as mitochondria, inhibits complex I and impairs oxidative phosphorylation.

2.3 Maneb

2.3.1 Background and history of use

Maneb, a member of the ethylene bisdithiocarbamate (EBDC) group of fungicides, was first registered for use on both food and ornamental crops (EPA, 2005). These fungicides are used to protect crops in the field as well as to prevent them from deterioration during transportation and storage. The EBDC group includes mancozeb and metiram, and all three compounds degrade into ethylenethiourea (ETU). The EPA considers risk from ETU derived from maneb, mancozeb and metiram (EPA, 2005). In 1992 the EPA cancelled the use of EBDCs on a variety of fruits and vegetables. Despite these restrictions, approximately 2.5 million pounds of maneb are still used annually on fruit, vegetable and nut crops. There is no residential use and careful agricultural practices ensure that there is no risk of residential exposures to maneb. Lettuce, almonds, peppers and walnuts are the main crops treated with maneb, and the risks of exposure to this pesticide as of 2005 warranted its reduction or outright cancellation for use on certain crops (EPA, 2005).

2.3.2 Toxic effects and mechanism of action

Overall, maneb is considered only moderately toxic to humans (Occupational Health Services, 1991) and little is known about its mechanism of action. Maneb increased cellular glutathione in SH-SY5Y cells and produced no reactive oxygen species (Roede et al., 2011). Both maneb and its relative, mancozeb led to mitochondrial dysfunction and reduced ATP levels in rat mesencephalic cells (Domico et al., 2006). Maneb and mancozeb reduced active respiration linked to NADH (Domico et al., 2006) while manganese-EBDC (the major component of maneb) appeared to specifically inhibit mitochondrial complex III (Zhang et al., 2003). Interestingly, while maneb and mancozeb were both toxic to mesencephalic neurons, their metabolite ETU was not (Domico et al., 2006), indicating that in terms of neurotoxicity, it is the primary components of the EBDC pesticides that are most toxic to neurons, and not their product. Interestingly, the mitochondrial dysfunction caused by maneb led to a reduction in ATP, which has a subsequent impact on the functioning of the ubiquitin-proteasome system. This system is involved in the intracellular degradation of proteins, the failure of which leads to the collection of protein aggregates and is associated with PD (McNaught et al., 2001). Manganese-EBDC was neurotoxic and led to reduced proteasome activity in a dopaminergic neuronal cell line (Zhou et al., 2004), which led to cytoplasmic inclusions containing α-synuclein, a hallmark of PD. Dopaminergic cells exhibited increased oxidative stress and neurotoxicity in response to manganese-EBDC. Increasing antioxidant levels via acetylcystein increased cell viability and eliminated the manganese-EBDC-induced increase in oxidative stress (Zhou et al., 2004).

Maneb has the interesting ability to potentiate the toxicity of other chemicals. Behaviorally, maneb intensified motor deficits experienced by mice treated with MPTP (Takahashi et al., 1989; Thiruchelvam et al., 2000b). The herbicide paraquat has a chemical structure very similar to that of MPP+, the active metabolite of MPTP. Thus, paraquat and maneb have been investigated for their combined ability to cause parkinsonian symptoms (Liou et al., 1997; Thiruchelvam et al., 2000a, 2000b). In a Taiwanese case study, PD risk was increased in those exposed to both paraquat and maneb (Liou et al., 1997). Exposure to this combination consistently led to significant changes in the nigrostriatal system (Thiruchelvam et al., 2000a, 2000b). In the presence of manganese-EBDC, striatal synaptosomes accumulated significantly more dopamine (Barlow et al., 2003). Toxicokinetic studies showed that manganese-EBDC given concurrently with paraquat led to more paraquat uptake by the brain than by other organs, indicating that the neurotoxicity of paraquat may be enhanced via maneb by directing more paraquat to be sequestered in the brain (Barlow et al., 2003).

It seems that exposure to both pesticides changes the mechanism of toxicity, at least as one study has shown. When tested independently, both paraquat and maneb triggered neuronal apoptosis via Bak, a pro-apoptotic member of the Bcl-2 gene family; however, when maneb and paraquat were tested together, Bak was inhibited and apoptosis was induced by the Bcl-2 member Bax (Fei & Ethell, 2008).

2.3.3 Development of animal models

Though there is evidence suggesting that exposure to the herbicide paraquat can contribute to PD, this model fails to take into account the overlap of areas exposed to other agrichemicals in addition to paraquat. This has led to the development of a mouse model of neurodegenerative disease using both maneb and paraquat to simulate exposure to both of these pesticides. In this model, male C57BL/6 mice (6 weeks old) were injected intraperitoneally with paraquat alone, maneb alone, or both paraquat and maneb (Thiruchelvam et al., 2000a, 2000b). The number of injections ranged from 4 (1 injection per week for 4 weeks; Thiruchelvam et al., 2000a) to 18 (2x per week for 9 weeks; Thiruchelvam et al., 2005). Essentially, the combined treatment with maneb and paraquat led to exacerbated effects on behavior, the dopaminergic system and the nigrostriatal system. Animals exhibited reduced locomotion, altered levels of dopamine and its metabolites, reduced dopamine transporter density and reduced number of dopaminergic neurons in the substantia nigra and striatum (Thiruchelvam et al., 2000a, 2000b). Unlike some other PD animal models, the maneb + paraquat model consistently induces a phenotype with impairments to both the motor system and nigrostriatal system.

The model has also been tested in a developmental context, where the mice were injected daily during post-natal days 5 – 19, and re-challenged with pesticide injection at 6.5 months of age to determine how early exposure may effect subsequent exposure (Thiruchelvam et al., 2002). Overall, the authors found that exposing mice at a young age to paraquat and maneb made them more susceptible to pesticides as adults, as only those exposed post-natally experienced toxic effects to all three treatments as adults (e.g. paraquat, maneb, and paraquat + maneb). Additionally, mice exposed to the two pesticides early in development showed more deficits than mice exposed to single pesticides (Thiruchelvam et al., 2002). It appears as though the maneb + paraquat model has no benefit over paraquat alone in rats (Cicchetti et al., 2005; Xu et al., 2011).

The combined paraquat and maneb system has been useful for investigating different neuroprotective or therapeutic agents, which may be helpful for people suffering from PD. Pre-treatment with lithium in the animal's food was able to eliminate α-synuclein protein aggregation and was neuroprotective against paraquat + maneb treatment of α-synuclein overexpressing transgenic mice (Kim et al., 2011). Similar neuroprotective effects were seen when mice exposed to maneb + paraquat were first treated with the naturally-occurring antioxidant silymarin or melatonin. Both silymarin and melatonin ameliorated the motor deficits induced by paraquat + maneb treatment such that these animals performed as well as control animals (Singhal et al., 2011). This striking effect may have been a consequence of silymarin's and melatonin's neuroprotective natures, as mice given one or the other prior to treatment with paraquat + maneb showed no significant loss of dopaminergic neurons in the substantia nigra (Singhal et al., 2011). Caffeine was also neuroprotective in the model, though the authors did not evaluate locomotor activity (Kachroo et al., 2010). Extract from the *Polygonum multiflorum* (PM) plant has also shown to be protective in this model. This herb has a long history in traditional Chinese medicine, and when given prior to and during exposure to maneb + paraquat, it was able to ameliorate the pesticide-induced motor deficits, reduction in striatal dopamine concentration and dopaminergic cell death (Li et al., 2005).

2.4 Endosulfan

2.4.1 Background and history of use

Endosulfan   (6,7,8,9,10,10-hexachloro-1,5,5a,6,9,9a-hexahydro-6,9-methano-2,4,3-benzadio-xathiepin 3-oxide) is a broad-spectrum organochlorine contact insecticide registered for use on fruits, vegetables, cereal grains, cotton and ornamental trees and shrubs. Endosulfan was first patented in 1956 and its use is an estimated 1.38 million pounds of active ingredient on average from 1987 – 1997 (EPA, 2002).

Endosulfan has established toxic effects, is persistent in the environment and has the ability to bioaccumulate (Chopra et al., 2011). Endosulfan is being phased out of use in many countries including the United States (EPA, 2010) and Canada (Health Canada, 2011). At the 5th meeting of the Conference of the Parties to the Stockholm Convention in 2011, endosulfan was banned globally. Strong opposition to this ban initially came from India, which uses and exports more endosulfan than any other country, with an estimated 113 kilotons used from 1958 – 2000 (Government of Canada, 2009). After a few concessions, India eventually agreed to the ban. The case of endosulfan use in India has a long and sordid history. Endosulfan has been applied to cashew crops via aerial spraying for over two decades. For years, residents of several villages in the Kasargod district of Kerala state had noticed increased illness and the death of animals, both wild and domestic, which seemed to coincide with the spraying (Adithya, 2009). Concerns of the villagers and doctors over the high rates of disease – including neurological disease, reproductive impairment, developmental problems and cancer – went largely unheard. Between 1998 and 2002, several groups, both national and international, conducted studies and concluded that endosulfan poisoning was responsible for the problems in Kasargod (Adithya, 2009; Quijano, 2002; Sivaraman et al., 2003). Doctors are now reporting a reduction in new cases since endosulfan was banned in Kerala in 2003 (Adithya, 2009).

2.4.2 Toxic effects and mechanism of action

Endosulfan is in the same general class of pesticides as DDT, the organochlorine pesticides. The very mechanism of action of these pesticides – which is neurotoxicity – makes them toxic to all animals, not just the insect pests that they target. They cause disruptions to the neuronal membrane which in turn lead to altered sodium and potassium kinetics (Hays & Laws, 1991). The result means that the sodium channel is not inactivated as normal, leading to a prolonged action potential (Hong et al., 1986; Soderlund & Bloomquist, 1989). This is proposed to be the mechanism leading to seizures seen in endosulfan poisoning (e.g. Pradhan et al., 1997). Additionally, endosulfan and related compounds are antagonists of the chloride channel linked to the receptor for gamma-aminobutyric acid (GABA), the major inhibitory neurotransmitter in the brain, leading to over-excitation of neurons (Cole & Casida, 1986; Klaassen, 1996).

There have been several recorded cases of acute accidental and non-accidental endosulfan poisoning in humans. In these cases, nearly all victims presented with seizure or convulsions, along with nausea and vomiting (Boereboom et al., 1998; Chugh et al., 1998; Karatas et al., 2006; Pradhan et al., 1997). In the case of 23 accidental poisonings (Karatas et al., 2006), all 23 people survived the insult. In one case of non-accidental poisoning, the patient died four days following exposure, the cause of death being cerebral edema (Boereboom et al., 1998). In a second case of non-accidental ingestion, the patient survived and appeared to be back to normal at a 3 month follow up (Pradhan et al., 1997). In a retrospective study of 52 cases of endosulfan intoxication, ingestion of the pesticide in excess of 35 g was the variable most likely to predict mortality (Moon & Chun, 2009).

A variety of *in vivo* studies have demonstrated the effects of endosulfan on the central nervous system. Rats treated orally with endosulfan showed behavioral deficits such as an increased time to learn and retain a task in an operant learning paradigm (Lakshmana & Raju, 1994), which may be related to alterations in neurotransmitter levels (Ansari et al., 1987, Lakshmana & Raju, 1994). Increased serotonin levels following endosulfan were thought to lead to a motivational deficit in rats, since the animals showed problems with memory tasks but not motor tasks (Paul et al., 1994). Reduced GABA levels were detected in the offspring of female rats fed endosulfan during pregnancy (Cabaleiro et al., 2008), and neonatal exposure also led to an increase in shock-induced aggression (Zaidi et al., 1985). Male mice injected with endosulfan as juveniles and subsequently challenged as adults showed reduced concentrations of dopamine and its metabolite DOPAC in brain samples (Jia & Misra, 2007).

There is a wealth of literature investigating the effects of endosulfan on the GABAergic system, and the ability of endosulfan to cause increased excitation in neurons is indicative of its neurotoxic nature. Fewer studies have directly investigated the toxic effects on neurons, either in culture or *in vivo*. The EPA's Federal Insecticide, Fungicide and Rodenticide Act guideline neurotoxicity studies failed to find overt signs of neuropathy associated with endosulfan treatment of rats (Silva & Gammon, 2009). In rats, endosulfan led to engorged blood vessels in the meninges and cerebral hemorrhages (Singh et al., 2007). Endosulfan-treated rats were shown to have increased lipid peroxidation in cerebral tissue, a sign of damaging oxidative stress (Hincal et al., 1995). *In vitro* studies have shown that endosulfan inhibited proliferation and differentiation of neural stem cells while also inhibiting neurite formation (Kang et al., 2001). Additionally, PC 12 cells incubated with endosulfan were reduced in number, had unusual morphology and exhibited increased apoptosis (Yang et al., 2004). Although studies directly showing the ability of endosulfan to kill neurons are lacking, what is clearly known is that endosulfan interferes with the GABAergic system, causing over-excitation due to lack of GABA receptor-mediated inhibition. It is known that indirect excitotoxicity of this type leads to cell death in some pathologies, such as brain damage associated with chronic alcoholism (Dodd, 2002).

2.4.3 Development of animal models

In the case of endosulfan, animal models have been restricted to establishing the typical toxicological endpoints such as LD_{50} levels. It is clear that endosulfan has a toxic effect on the brain, but there is no clearly established animal model developed in order to evaluate its role specifically in neurodegenerative disease. In fact, there are relatively few *in vitro* studies showing that endosulfan is capable of killing neurons. Given the global ban on endosulfan, the establishment of an animal model for continued neurotoxicity testing seems somewhat unnecessary, except for studying the long-term effects of endosulfan exposure, as exposed populations still exist. Given the number and variety of pesticides that human populations are exposed to regularly, it seems unlikely that a banned substance will continue to be researched in the way that endosulfan has in the past.

2.5 Atrazine

2.5.1 Background and history of use

Atrazine (6-chloro-N^2-eyhyl-N^4-isopropyl-1,3,5-triazine-2,4-diamine) is one of the most heavily used herbicides in the United States, with an estimated use of 76.5 million pounds per year (EPA, 2003). Atrazine was first registered for use as an herbicide by JR Geigy SA (currently known as Syngenta) in 1958, and is widely employed in the control of broadleaf plants (EPA, 2003; Gammon et al., 2005). Measures were taken in the 1990s to reduce the amount of atrazine contamination in surface and groundwater, including the establishment of a Maximum Contaminant Level of 3 parts per billion for atrazine (EPA, 2003). Despite established usage limits, data show that the overall use of atrazine has changed very little since the late 1980s (Kiely et al., 2004). Approximately 80 million pounds of atrazine are applied annually (Kiely et al., 2004), with particularly heavy usage on corn, sorghum and sugarcane. In fact, atrazine is the main pesticide used on these three crops, and treatment with atrazine amounts to 75% of all American corn, 58.5% of all sorghum and 76% of all sugarcane (EPA, 2003). In addition to its pervasive use in agriculture, atrazine is also used in residential settings and on golf courses, adding to the potential for human exposure, though data on these routes of exposure are lacking. Oral exposure through food consumption of atrazine is very low (Ribaudo & Bouzaher, 1994), but the chemical properties of atrazine mean that it enters surface and groundwater through leaching and runoff (Ribaudo & Bouzaher, 1994). Exposure through drinking water is the most likely route of exposure for the majority of the American population, except for occupational exposure experienced by atrazine applicators and farmers and their families, who are at an increased risk of exposure (Curwin et al., 2007). A complete European ban on atrazine came into effect between 2005 - 2007 (Ackerman, 2007). Given the estimated economic impacts of banning atrazine in the United States, it is unlikely that total or even partial bans on its use are forthcoming

(Ribaudo & Bouzaher, 1994). Whether or not these estimated impacts are realistic is another question (Ackerman, 2007), and the use of atrazine in the United States remains a contentious issue.

2.5.2 Toxic effects and mechanism of action

Atrazine functions as an herbicide by inhibiting photosynthesis in the target plant. Specifically, atrazine prevents electron transfer at complex II in the chloroplast (Gysin & Knuesli, 1960). Thus, the mechanism of toxicity is similar to the mechanism whereby dopaminergic neurons are killed in the paraquat model of PD (Franco et al., 2010).

The possible neurotoxic effects of atrazine are much less researched than the endocrine effects, which have received much attention. Despite the lower number of studies, there is evidence that atrazine is neurotoxic. Atrazine administered orally is capable of crossing the blood-brain barrier and could be measured in brain tissue, along with didealkyl atrazine, its major metabolite (Ross et al., 2009). A small body of literature has examined the neurotoxic effects of atrazine in both *in vitro* and *in vivo* studies. Many of these studies show damage which is similar to that seen in neurodegenerative diseases such as PD. *In vivo*, both chronic and acute exposures to atrazine affect brain monoamine systems with associated changes to behavior. Behavioral effects vary, however, such that dietary atrazine treatment led to hyperactivity after 6 months of treatment (Rodríguez et al., 2005), or hypoactivity after 8 months of treatment (Bardullas et al., 2011). Different strains used (Long-Evans versus Sprague-Dawley) might account for these results, or the fact that testing continued throughout the year, which might have affected activity levels overall (Bardullas et al., 2011). Atrazine treatment also impaired performance in learning tasks (Bardullas et al., 2011).

The *in vivo* chronic treatment of rats with atrazine also leads to alterations in brain neurotransmitter levels. After 6 or 12 months of atrazine, dopamine levels were reduced in the striatum (Bardullas et al., 2011; Rodriguez et al., 2005), and norepinephrine levels were reduced in the prefrontal cortex (Rodriguez et al., 2005). In fact, dopamine and its metabolites were reduced in the murine striatum after as few as 14 days of atrazine exposure (Coban & Filipov, 2007). After 6 months, serotonin was reduced in the hypothalamus (Rodriguez et al., 2005), and in the striatum after 14 days (Coban & Filipov, 2007), while reduced serotonin was not seen after 12 months in a separate experiment (Bardullas et al., 2011). Atrazine treatment reduced numbers of tyrosine hydroxylase immunoreactive neurons in the substantia nigra pars compacta and the ventral tegmental area (Coban & Filipov, 2007; Rodriguez et al., 2005), indicating that it is neurotoxic to dopaminergic neurons. Atrazine also caused general neurodegeneration in the hippocampus of the female mouse (Giusi et al., 2006). *In vivo* microdialysis experiments further showed that dopamine release in the striatum is reduced as a consequence of an acute exposure to atrazine (Rodríguez et al., 2005).

In vitro studies show that exposure to atrazine disrupts the dopaminergic system. Studies using striatal slices showed that atrazine treatment lowered the level of dopamine released in the striatum, but it did not affect the level of the rate-limiting enzyme tyrosine hydroxylase. There is an indication that the reduction in dopamine could be associated with a decrease in dopamine-producing neurons in the substantia nigra (Filipov et al., 2007).

Additionally, the uptake and sequestration of dopamine into vesicles appeared impaired and perhaps as a consequence of this, the ratio of dopamine metabolites/dopamine increased (Filipov et al., 2007; Hossain & Filipov, 2008). PC12 cells are neuronal cells derived from a tumour of the rat adrenal medulla that produce catecholamines including dopamine. Intracellular dopamine concentration was reduced in undifferentiated PC12 cells treated with atrazine, and the relationship was dose-dependent, while tyrosine hydroxylase levels, though slightly reduced, were not significantly effected (Das et al., 2000; Das et al., 2003). Intracellular norepinephrine was decreased and the cells exhibited reduced norepinephrine release and reduced dopamine β-hydroxylase, the enzyme which converts dopamine to norepinephrine (Das et al., 2000; Das et al., 2003). Moreover, atrazine metabolites produced unique and varying effects on dopamine and norepinephrine release in PC12 cells, with certain metabolites increasing levels of dopamine and/or norepinephrine, and other decreasing catecholamine levels (Das et al., 2001).

2.5.3 Development of animal models

Unlike other herbicides, a standard model for testing the neurodegenerative effects of atrazine has not been established. The chemical has been tested on rats and mice *in vivo*, but the protocols vary in terms of route of administration, dose, time, animal strain and endpoint. Before atrazine can be firmly established as a confirmed neurotoxicant, these parameters need to be standardized. A lack of standardization across studies may help to account for differences seen in results, and though these studies point to atrazine as being a toxic agent, the link remains tenuous. Behavioral endpoints need to be established to help further demonstrate the detrimental effects of atrazine. It has been demonstrated in cell culture and brain slice preparations that atrazine negatively impacts the catecholaminergic system, but the long-term ramifications of this toxicity are not clear, and need to be known in order to say with more confidence that atrazine is linked to neurodegenerative disease.

2.6 Aldrin/dieldrin

2.6.1 Background and history of use

Aldrin is an organochlorine insecticide which breaks down into dieldrin. It is a PBT chemical, meaning that it is persistent, bioaccumulative and toxic. Aldrin was widely used in the United States to control insect infestations on corn, cotton and citrus until it was completely banned for agricultural use in 1985 amid health concerns (EPA Persistent Bioaccumulative and Toxic (PBT) Chemical Program, 2011). However, given its persistent nature, aldrin is still found in the environment where humans and animals may still be at risk of exposure. Since the majority of the literature concerns specifically dieldrin, it is dieldrin that will primarily be covered here.

2.6.2 Toxic effects and mechanism of action

As an organochlorine insecticide, dieldrin is similar in action to endosulfan. It kills insects via the same mechanism as endosulfan, by inhibiting GABA receptors and causing neuronal over-excitation (e.g. Ikeda et al., 1998). Dieldrin is neurotoxic, particularly to dopaminergic cells (reviewed by Kanthasamy et al., 2005). Dieldrin produced time- and dose-dependent increases in neurotoxicity in both GABAergic and dopaminergic cells in mesencephalic cell culture, but the effect was more pronounced in the dopaminergic neurons (Sanchez-Ramos et al., 1998). Moreover, dieldrin was toxic to dopaminergic PC12 cells at lower doses than it was to both pancreatic endocrine cells and human cortical neurons (Kitazawa et al., 2001). Dieldrin may be toxic to dopaminergic neurons by producing reactive oxygen species (ROS). Treatment of dopaminergic SN4741 cells with dieldrin lead to a slow increase in ROS production, particularly H_2O_2, which eventually led to apoptosis (Chun et al., 2001). ROS combines with nitric oxide, producing reactive nitrogen species which affect the respiratory chain of the mitochondria resulting in a decrease in ATP synthesis and increase in lipid peroxidation (Ebadi & Sharma, 2003). These changes in the membrane properties affect cellular homeostasis leading to dysfunction of the mitochondria and neurotoxicity.

In addition to killing dopaminergic neurons, dieldrin may also cause dopamine dysfunction. While Thiffault and colleagues (2001) found that treating mice with a single injection of dieldrin was insufficient to cause any reduction in dopamine or its metabolites in the striatum, long term exposures have the potential to severely alter normal brain chemistry. Mice exposed for 30 days showed decreases in dopamine metabolite levels, as well as decreases is dopamine transporter expression in the striatum (Hatcher et al., 2007). Developmental exposure to dieldrin appears to be detrimental as well, as female mice exposed to low diedrin levels had offspring that were predisposed to increased toxicity from other agents (Richardson et al., 2006). Additionally, at 12 weeks of age, mice exposed *in utero* showed dose-dependent increases in dopamine transporter and vesicular monoamine transporter 2 mRNA levels, indicative of dopamine dysfunction, and opposite to what was seen with post-natal exposures.

Dieldrin produces contrary effects on dopamine efflux in PC12 cells. Extracellular dopamine was increased while intracellular levels were decreased when dopamine content was measured via HPLC (Kitazawa et al., 2001). When measured in PC12 cells loaded with radionuclide-labeled dopamine, dopamine efflux was found to be inhibited in the presence of dieldrin (Alyea & Watson, 2009). Time course and other methodological differences could account for these contradictory results.

2.6.3 Development of animal models

As is the case with several of the pesticides mentioned above, there has been little progress in terms of the formal development of an animal model for investigating the neurotoxic effects of dieldrin. Although dieldrin has often been administered directly to mice (Hatcher et al., 2007; Richardson et al., 2006; Thiffault et al., 2001) these studies have used varying treatment parameters, meaning that often conflicting results must be evaluated carefully. The ban on dieldrin also means that there is less interest in fully establishing its status as a factor in neurodegeneration, and it is unlikely to see further development as a model of PD. However, dieldrin still persists in the environment, making low-dose chronic exposure studies all the more important. Organochlorines appear to be particularly high risk factors for the development of neurodegenerative disease (Seidler et al., 1996) and the toxicity of dieldrin, endosulfan and DDT are not in dispute. The development of a standardized animal model for testing dieldrin will not only help to further identify its particular neurotoxicity, but may also help contribute to our overall understanding of how the organochlorines may contribute to PD development.

3. Evidence for a correlation of herbicide exposure and neurodegenerative disease

Animal models of pesticide contribution to neurodegenerative disease are largely focussed on the development of Parkinson's disease. These have been discussed in the above section. For the purpose of reviewing additional evidence for correlation of pesticide exposure and the development of neurodegenerative disease, this section will discuss the human epidemiological evidence. This section will further be divided into reviews of the neurodegenerative diseases most often associated with pesticide exposure: Parkinson's disease, Alzheimer's disease and Amyotrophic lateral sclerosis.

3.1 Parkinson's disease (PD)

Interest in the association of PD and pesticide exposure by sparked with the observation nearly 30 years ago that exposure to the chemical MPTP caused PD symptoms (Langston et al., 1983). MPP+, the toxic metabolite of MPTP which is structurally similar to paraquat, is selectively uptaken by dopaminergic neurons. Recently, the epidemiology of PD has been extensively reviewed, including thorough reviews of the association between PD and pesticides (Brown et al., 2006; Wirdefeldt et al., 2011). The epidemiological data for pesticide exposure and PD are extensive, and the following review is not meant to be exhaustive.

In examining the case-controlled literature, Wirdefeldt and colleagues found that there were 38 published case-control studies looking at pesticide exposure and incidence of PD. Of these, exactly half found a positive association, and half found no association (Wirdefeldt et al., 2011). When 31 of the studies were presented in a forest plot, it was clear that although half of the 38 case-controlled studies found no significant association, most of the studies still report increased risk for the case group (Brown et al., 2006). When these studies were further broken down into those which attempted to examine the effects of different types of pesticides (herbicides, fungicides, insecticides and organochlorines), herbicide exposure was positively associated with PD, as was insecticide exposure (Brown et al., 2006). In one study, after controlling for additional pesticide exposure, herbicide exposure was still a significant independent risk factor for PD (Semchuk et al., 1992). Paraquat exposure specifically presented a significant risk (Hertzman et al., 1990; Liou et al., 1997), and risk was increased, though non-significantly, in other studies (Firestone et al., 2005; Hertzman et al., 1994). Overall, this indicated that there is a link between Paraquat exposure and the incidence of PD. In a meta-analysis of 19 studies, the authors calculated a combined odds ratio for PD risk of 1.94, indicating a positive association (Priyadarshi et al., 2000).

There appear to be several factors influencing the possibly that an individual will develop PD as a consequence of exposure to agrichemicals. Studies indicate that there may be a critical period of exposure, as greater risk was associated with exposure at specific age periods (Semchuk et al., 1992). Perhaps exposure duration is a more important factor, as several studies indicate that risk increases with an increasing duration of exposure, particularly in excess of 10 or 20 years (Gorell et al., 1998; Liou et al., 1997; Seidler et al., 1996). As with some other toxins, the danger may be in the dose, as there was a greater association seen with high doses compared to low doses (Nelson et al., 2000, as cited in Brown et al., 2006) although trying to determine dose in case-controlled studies is often tricky and imprecise, and in other cases, the relationship between dose and risk may even be reversed (Kuopio et al., 1999). Most likely, it is a combination of long duration and high dose that has a greater effect on risk (Nelson et al., 2000, as cited in Brown et al., 2006; Seidler et al., 1996).

PD risk is also associated with different – but related – environmental factors such as rural living, farming and drinking well water. Studies have shown higher occurrence of PD in rural versus urban areas in Canada (Barbeau et al., 1987), the United States (Lee et al., 2002), and Denmark (Tuchsen et al., 2000). Drinking well water is a risk factor for PD (Tsai et al., 2002); however, this tends to be dependent on rural living, so the two factors are interrelated (Koller et al., 1990). If farming is a legitimate risk factor for PD, it is difficult to detect in case-control studies. Wirdefeldt and colleagues (2011) report that of 34 case-control studies that considered farming and PD prevalence, there was significantly increased risk in only 7 studies. A meta-analysis published by Priyadarshi and colleagues (2001) found an association with farming in the United States (including exposure to farm animals or living on a farm). The meta-analysis is of interest because the authors calculated odds ratios based on all the existing published data, rather than just grouping the studies together based on whether or not the original authors found a significant association.

Several confounding factors need to be addressed in these studies. For example, smoking is negatively correlated with PD incidence, but very few case-control studies have corrected for smoking. Defining pesticide exposure is challenging, and considering this, it is not surprising that few studies manage to find a significant association. Exposure assessment largely relies on the ability of the patient to recall their exposure based on general questions, and often these questions aren't given in the published study. Exposure is often defined in a dichotomous way. Where exposures are residential or non-occupational, patient recall may be unreliable. Clearly there needs to be a more reliable and objective way of establishing pesticide exposure. Specific biomarkers could be used, or tissue samples could be analyzed for pesticide concentrations. For example, higher levels of dieldrin (Corrigan et al., 1998) and lindane (Corrigan et al., 2000) were detected in brain samples taken from PD patients post-mortem. Increasing plasma dieldrin levels were also associated with PD in never smokers (Weisskopf et al., 2010). For a persistent organic pollutant such as dieldrin, this is simple enough, but it may be troublesome for pesticides which break down more rapidly, or do not accumulate.

All of us are exposed to pesticides and environmental toxins on a regular basis and the incidence of PD does not seem to reflect this fact. Humans have evolved biological mechanisms for the detoxification and metabolism of xenobiotics, and it is feasible that perturbations in these mechanisms could help determine whether or not an individual develops a neurodegenerative disease in response to pesticides. Individuals with mutations in the CYP2D6 gene – which is responsible for metabolizing environmental toxicants – may lack sufficient means to remove toxins before they contribute to neurodegeneration. These individuals have been termed "poor metabolizers" and show undetectable CYP2D6 activity (reviewed by Elbaz et al., 2007). There appears to be a strong association between CYP2D6 dysfunction and PD occurrence.

Overall, the evidence suggests an association between pesticide exposure and PD risk. Brown and colleagues (2006) demonstrated that this may be particularly true for herbicide exposure. Future epidemiological studies need to include certain methodological considerations, chiefly exposure assessment in non-occupational situations. Additionally, studies need to accurately assess additional factors such as smoking and control for these factors in statistical analyses.

3.2 Alzheimer's disease (AD)

There is little evidence for the role of pesticide exposure in the development of AD, the most common type of dementia. As with PD, rural living is associated with higher incidence rates of AD in countries including Italy (Rocca et al., 1990) and Finland (Sulkava et al., 1988). A recent study conducted in the Andalusia region of Spain found that there was a greater risk of developing AD in districts with high pesticide use (Parrón et al., 2011).

There are two case studies in the literature which describe individuals aged 55 (Cannas et al., 1992) and 59 (Lake et al., 2004) diagnosed with AD who experienced long-term occupational pesticide exposure. These case studies offer little evidence that chronic pesticide exposure can lead to AD, but occupational exposure is a legitimate risk factor. A study investigating neurodegenerative disease incidence in different occupations found increased AD (and PD) in non-horticultural farmers below the age of 65 (Park et al., 2005). In a cohort study of the elderly in France, occupational exposure to pesticides decreased cognitive performance and increased risk of developing both AD and PD in men (Baldi et al., 2003). A study of the residents of an agricultural community in Utah found increased risk of AD and dementia among those exposed to pesticides, with organophosphate pesticides being identified as particularly dangerous (Hayden et al., 2010), while a case-control study conducted in Quebec, Canada found no significant risk of AD with exposure (Gauthier et al., 2001). However, an analysis of the Canadian Study of Health and Aging looking at 258 cases of clinically diagnosed AD found a positive association with pesticides (Canadian Study on Heath and Aging Working Group, 1994). Compared to PD, there is more consensus that pesticide exposure increases risk of AD. Although there have been very few studies, most have found a significant association.

3.3 Amyotrophic lateral sclerosis (ALS)

ALS is a motor neuron disease caused by the degeneration of neurons in the brain and spinal cord. As with AD, there is little scientific literature on the correlation between pesticides and development of ALS, although from what little data are available, there does appear to be a relationship. Of 6 identified epidemiological studies that included pesticide risk of ALS, 5 of these found a significant association. Most recently, in a case-control study of ALS in northern Italy, Bonvicini and colleagues (2010) found that compared to age- and sex-matched controls, more ALS cases had experienced occupational pesticide exposure in excess of 6 months. There was an increased risk of ALS among employees of the Dow Chemical Company who were exposed to the herbicide 2,4-dichlorophenoxyacetic acid versus other Dow employees (Burns et al., 2001). Similar studies have found positive associations for populations exposed to pesticides (McGuire et al., 1997; Qureshi et al., 2006), and the relationship was found to be dependent on dose (Morahan & Pamphlett, 2006). Overall, the relationship tends to be stronger for males than females (McGuire et al., 1997; Morahan & Pamphlett, 2006). In one study using participants in the American Cancer

Society's Cancer Prevention Study II, no association between pesticide exposure and ALS was detected (Weisskopf et al., 2009).

Although epidemiological studies are lacking, there are other hypotheses regarding how pesticide exposures may contribute to ALS. Evidence suggests that certain individuals are genetically susceptible to ALS due to mutations in specific genes which are involved in the handling and detoxification of xenobiotics. The paraoxonase enzymes are responsible for detoxifying organophosphates and are coded for by the genes PON1-3. These genes have been implicated in the development of ALS through pesticide exposure. When investigating PON polymorphisms in a control population versus a population exposed to organophosphates, Sirivarasai and colleagues (2007) found that the exposed group exhibited three polymorphisms in PON1 that were not observed in the control group. The exposed group also had reduced enzyme levels (Sirivarasai et al., 2007). A number of studies have found polymorphisms in the PON locus to be positively associated with ALS (Cronin et al., 2007; Landers et al., 2008; Saeed et al., 2006; Slowik et al., 2006; Valdmanis et al., 2008). However, the data are far from conclusive, and a recent meta-analysis failed to find a significant association between ALS and PON (Wills et al., 2009).

In conclusion, given the evidence presented above, occupational exposure to pesticides appears to present a significant risk for developing various neurodegenerative diseases. Certain allelic mutations may pre-dispose individual to increased toxicity from pesticides if they lack sufficient detoxification mechanisms. Case-control studies have proven useful for establishing links between exposure and disease development, but certain methodological considerations must be addressed. Studies must control for certain confounding factors such a smoking, and pesticide exposure assessment should be standardized and refined.

4. Recent advances in the development of neuroprotective agents

Neurons are post-mitotic, that is they do not undergo cell division. Hence, populations of neurons that have been killed cannot be easily replaced by surviving cells. These cells are also more susceptable to oxidative stress in comparison to other cell types. Morevoer, there is an increase in oxidative stress and hence oxidative damage with age. In PD, the patient remains asymptomatic until 50% of the neurons in the substantia nigra are lost. No drug designed so far can prevent neurodegeneration and halt the progression of the disease. The primary treatment available for PD is administration of dopamine agonists which inhibit dopamine degradation and provide temporary symptomatic relief (Obeso et al., 2000). The use of antioxidants such as Vitamin E has failed for PD, though it has shown more positive results for diseases such as AD and ALS (Sano et al., 1997; The Parkinson's Study Group, 1993).

Levodopa – L-3, 4-dehydroxyphenylalanine is a standard, highly effective drug used for the treatment of PD and other therapies that are being studied are usually compared to it (Fahn, 1999). Upon crossing the blood brain barrier , levodopa is converted to dopamine by the enzyme L- aminodecarboxylase. When levodopa is converted to dopamine, dopamine receptors are activated which in turn lessens the symptoms of PD. Levodopa is more effective in higher doses in comparison with lower doses; however, the high dose might result in adverse effects such as dyskinesia (Fahn et al., 2004).

Levodopa does not cure PD and the dose might have to be standardised for each patient.

The substantia nigra region of the brain is prone to high oxidative stress even under normal conditions because of the high levels of reactive metabolites produced by dopamine. Using neuroprotective agents that reduce oxidative stress and protect the mitochondria, thereby decreasing neurodegeneration, would be the ideal therapy for PD. Coenzyme Q10 (CoQ10), a naturally occurring compound that participates in the electron transport chain, is an effective antioxidant. Both fat and water soluble CoQ10 have been used for animal and preclinical studies. The clinical trial for fat soluble CoQ10 formulation in the treatment of PD has been halted as it was not effective. The water soluble formulation (www.zymes.com/) of CoQ10 (consisting of CoQ10, polyethylene glycol, and α- tocopherol) however, has been shown to be an effective prophylactic agent that prevents progression to PD. A dose of 50 μg/ml has shown to be effective in preventing the degeneration of dopaminergic neurons in Long Evans Hooded rats injected with paraquat (Somayajulu-Nitu et al., 2009). These rats did not show PD - like symptoms and there was no motor impairment or decrease in the number of dopaminergic neurons. Water soluble CoQ10 may also be useful as a therapeutic treatment for PD (Facecchia et al., unpublished data)

Pre-treatment of C56BL/6 mice with caffeine before injecting them with a combination of maneb and paraquat helps reduce the loss of dopaminergic neurons. A chronic dose of 20 mg/Kg increased the expression of tyrosine hydroxylase in the substantia nigra region (Kachroo et al., 2010).

Other neuroprotective agents that have been studied and have shown positive results are silymarin and melatonin (Singhal et al., 2010). Lithium feeding for a period of three months prevented accumulation of α- synuclein in nine - month old α- synuclein transgenic mice that were injected with paraquat and maneb. It also protected α-synuclein enhanced green fluorescent protein overexpressing dopaminergic N27 cells that were treated with hydrogen peroxide (Kim et al., 2011).

Polygonum multiflorium is a Chinese herb and its extract contains an ethanol soluble fraction and an ethanol insoluble fraction. The ethanol soluble fraction has been shown to prevent a decrease in the number of dopaminergic neurons in the substantia nigra region of C57BL/6 mice injected with paraquat and maneb twice a week for six weeks (Li et al., 2005).

Autophagy is an inducible process that is activated by factors such as stress, pathogenic invasion, or nutrient or growth factor deprivation. Mutation of genes such as *a- synuclein* and *tau* could result in the expression of misfolded and aggregated proteins that can be removed by autophagy enhancement thereby preventing neurodegeneration (Pan et al., 2009). Feeding mice with 1% trehalose for 2.5 months has been shown to decrease PD - like symptoms in transgenic mice with mutation in the *tau* gene (Rodriguez- Navarro et al., 2010).

5. Conclusions and recommendations

In conclusion, we have provided evidence that environmental exposures to pesticides can contribute to the development of neurodegenerative diseases. Typical toxicological testing does not normally account for the neurotoxic effects of pesticides, rather, these tests focus on establishing LC50 values and evaluating toxicity to organs such as kidney, liver and lungs.

Neurotoxicity testing seldom evaluates the effects of long-term, low-dose exposure to pesticides, which is a much more likely human exposure scenario. We suggest that long-term low-dose *in vivo* and *in vitro* testing of pesticides for neurotoxicity should become standard procedure before introducing new products into the environment. Neurons are particularly susceptible to long-term exposure to toxic chemicals given their post-mitotic and non-regenerating nature. Neurotoxicology studies should examine biochemical, behavioral and histological effects of pesticide exposure in order to establish a more complete picture of the potential toxicity of chemicals. Moreover, behavioral testing should examine not only motor effects, but also deficits in learning, memory or cognition (Walsh & Chrobak, 1987), which may not be as overt but are still highly detrimental. Although mammalian models have long been the standard for toxicity testing, non-mammalian systems are becoming more common, and present new alternatives for long-term neurotoxicity testing (Peterson et al., 2008). Histological examination of the brain for signs of toxicity is obviously an important consideration. Assessing neurodegeneration in the dopaminergic system, particularly the substantia nigra, may be considered as a parameter for neurotoxicity testing, given the apparent susceptibility of the dopaminergic system to a variety of compounds, some pesticides included (Storch et al., 2004). Hopefully, future toxicity studies will consider the potential of a pesticide to induce cell death specifically in the brain, and probable neurotoxicants will be kept out of the environment.

6. Acknowledgements

The authors wish to thank Miss. Manika Gupta for critical reading of this manuscript and for providing valuable comments. We gratefully acknowledge funding from the Natural Science and Engineering Research Council of Canada, Canadian Institute for Health Research, and the Michael J. Fox Foundation NY.

7. References

Ackerman, F. (2007). The economics of atrazine. *International Journal of Occupational and Environmental Health*. Vol.13, No.4, (October-December 2007), pp. 441-449.

Adithya, P. (2009). India's endosulfan disaster. A review of the health impacts and status of remediation. Thanal. July 2011, Available from:
http://p7953.typo3server.info/panfiles/download/endosulfan-aditya.pdf

Alyea, RA., & Watson, CS. (2009). Differential regulation of dopamine transporter function and location by low concentrations of environmental estrogens and 17β-estradiol. *Environmental Health Perspectives*. Vol.117, No.5, (May 2009), pp. 778-783.

Ansari, RA., Husain, K., & Gupta, PK. (1987). Endosulfan toxicity influence on biogenic amines of rat brain. *Journal of Environmental Biology*. Vol.8, No.3, pp. 229-236.

Baldi, I., Lebailly, P., Mohammed-Brahim, B., Letenneur, L., Dartigues, JF., & Brochard, P. (2003). Neurodegenerative diseases and exposure to pesticides in the elderly. *American Journal of Epidemiology*. Vol.157, No.5, (March 2003), pp. 409-141.

Barbeau, A., Roy, M., Bernier, G., Campanella, G., & Paris, S. (1987). Ecogenetics of Parkinson's disease: prevalence and environmental aspects in rural areas. *Canadian Journal of Neurological Sciences*. Vol.14, No.1, (February 1987), pp. 36-41.

Bardullas, U., Giordano, M., & Rodríguez, VM. (2011). Chronic atrazine exposure causes disruption of the spontaneous locomotor activity and alters the striatal dopaminergic system of the male Sprague-Dawley rat. *Neurotoxicology and Teratology.* Vol.33, No.2, (March-April 2011), pp. 263-272.

Barlow, BK., Thiruchelvam, MJ., Bennice, L., Cory-Slechta, DA., Ballatori, N., & Richfield, EK. (2003). Increased synaptosomal dopamine content and brain concentration of paraquat produced by selective dithiocarbamates. *Journal of Neurochemistry.* Vol.85, No.4, (May 2003), pp. 1075-1086.

Betarbet, R., Sherer, T B., MacKenzie, G., Garcia-Osuna, M., Panov, A V., & Greenamyre, JT. (2000) Chronic systemic pesticide exposure reproduces features of Parkinson's disease. *Nature Neuroscience.* Vol.3, No.12, (December, 2000) pp.1301–1306.

Boereboom, FT. van Dijk, A., van Zoonen, P., & Meulenbelt, J. (1998). Nonaccidental endosulfan intoxication: a case report with toxicokinetic calculations and tissue concentrations. *Journal of Toxicology. Clinical Toxicology.* Vol.36, No4, pp. 345-352.

Bonvicini, F., Marcello, N., Mandrioli, J., Pietrini, V., & Vincenti, M. (2010). Exposure to pesticides and risk of amyotrophic lateral sclerosis: a population-based case-control study. *Annali dell'Istituto Superiore di Sanità.* Vol.46, No.3, pp. 284:287.

Brown, TP., Rumsby, PC., Capleton, AC., Rushton, L., & Levy, LS. (2006). Pesticides and Parkinson's disease--is there a link? *Environmental Health Perspectives.* Vol.114, No.2, (February 2006), pp. 156-64.

Burns, RS., Chineh, RS., Markey, SP., Ebert, MH., Jacobowitz, DM., Kopin, IJ. (1983). A primate model of parkinsonism: selective destruction of dopaminergic neurons in the pars compacta of the substantia nigra by N-methyl-4-phenyl-1,2,3,6-tetrahydropyridine. *PNAS.* Vol. 80, No. 14, (July, 1983) pp.4546-4550.

Burns, CJ., Beard, KK., & Cartmill, JB. (2001). Mortality in chemical workers potentially exposed to 2,4-dichlorophenoxyacetic acid (2,4-D) 1945-94: an update. *Occupational and Environmental Medicine.* Vol.58, No.1, (January 2001), pp. 24-30.

Cabaleiro, T., Caride, A., Romero, A., & Lafuente, A. (2008). Effects of in utero and lactational exposure to endosulfan in prefrontal cortex of male rats. *Toxicology Letters.* Vol.176, No.1, (January, 2008), pp. 58-67.

Castello, PR., Drechsel, DA., & Patel, M. (2007). Mitochondria are a major source of paraquat-induced reactive oxygen species production in the brain. *The Journal of Biological Chemistry.* Vol.282, (March, 2007), pp. 14186-14193.

Canadian Study of Health and Aging Working Group. (1994). The Canadian Study of Health and Aging: risk factors for Alzheimer's disease in Canada. *Neurology.* Vol.44, No.11, (November 1994), pp. 2073-2080.

Cannas, A., Costa, B., Tacconi, P., Pinna, L., & Fiaschi, A. (1992). Dementia of Alzheimer type (DAT) in a man chronically exposed to pesticides. *Acta Neurologica.* Vol.14, No3, (June 1992), pp. 220-223.

Chopra, AK., Sharma, MK., & Chamoli, S. (2011). Bioaccumulation of organochlorine pesticides in aquatic system--an overview. *Environmental Monitoring and Assessment.* Vol.173, No.1-4, (February 2011), pp. 905-916.

Chugh, SN., Dhawan, R., Agrawal, N., & Mahajan, SK. (1998). Endosulfan poisoning in Northern India: a report of 18 cases. *International Journal of Clinical Pharmacology and Therapeutics.* Vol.36, No.9, (September 1998), pp. 474-477.

Chun, HS., Gibson, GE., DeGiorgio, LA., Zhang, H., Kidd, VJ., & Son, JH. (2001). Dopaminergic cell death induced by MPP(+), oxidant and specific neurotoxicants shares the common molecular mechanism. *Journal of Neurochemisty.* Vol.76, Vo.4, (February 2001), pp. 1010–1021.

Cicchetti, F., Lapointe, N., Roberge-Tremblay, A., Saint-Pierre, M., Jimenez, L., Ficke, B.W., & Gross, RE. (2005). Systemic exposure to paraquat and maneb models early Parkinson's disease in young adult rats. *Neurobiology of Disease.* Vol.20, No2, (November 2005), pp. 360-371.

Coban, A., & Filipov, NM. (2007). Dopaminergic toxicity associated with oral exposure to the herbicide atrazine in juvenile male C57BL/6 mice. *Journal of Neurochemistry.* Vol.100, No.5, (March 2007), pp. 1177-1187.

Cole, LM., & Casida, JE. (1986). Polychlorocycloalkane insecticide-induced convulsions in mice in relation to disruption of the GABA-regulated chloride ionophore. *Life Sciences.* Vol.39, No.20, (November 1986), pp. 1855-1862.

Conning, DM., Fletcher, K., & Swann, ABB. (1969). Paraquat and related bipyridyls. *British Medical Bulletin.* Vol. 25, No. 3, pp.245-249.

Corrigan, FM., Murray, L., Wyatt, CL., & Shore, RF. (1998). Diorthosubstituted polychlorinated biphenyls in caudate nucleus in Parkinson's disease. Experimental Neurology. Vol.150, No.2, (April 1998), pp. 339–342.

Corrigan, FM., Wienburg, CL, Shore, RF., Daniel, SE., & Mann, D. (2000). Organochlorine insecticides in substantia nigra in Parkinson's disease. *Journal of Toxicology and Environmental Health. Part A.* Vol.59, No.4, (February 2000), pp. 229–234.

Cronin, S., Greenway, MJ., Prehn, JH., & Hardiman, O. (2007). Paraoxonase promoter and intronic variants modify risk of sporadic amyotrophic lateral sclerosis. *Journal of Neurology, Neurosurgery, and Psychiatry.* Vol.78, No.9, (September 2007), pp. 984-986.

Curwin, BD., Hein, MJ., Sanderson, WT., Striley, C., Heederik, D., Kromhout, H., Reynolds, SJ., & Alavanja, MC. (2007). Urinary pesticide concentrations among children, mothers and fathers living in farm and non-farm households in Iowa. *The Annals of Occupational Hygiene.* Vol.51, No.1, (January 2007), pp. 53-65.

Das, PC., McElroy, WK., & Cooper, RL. (2000). Differential modulation of catecholamines by chlorotriazine herbicides in phenochromocytoma (PC12) cells *in vitro. Toxicological Sciences.* Vol.56, No.2, (August 2000), pp. 324-331.

Das, PC., McElroy, WK., & Cooper, RL. (2001). Alteration of catecholamines in phenochromocytoma (PC12) cells *in vitro* by the metabolites of chlorotriazine herbicide. *Toxicological Sciences.* Vol.59, No.1, (January 2001), pp. 127-137.

Das, PC., McElroy, WK., & Cooper, RL. (2003). Potential mechanisms responsible for chlorotriazine-induced alterations in catecholamines in phenochromocytoma (PC12) cells. *Life Sciences.* Vol.73, No.24, (October 2003), pp. 3123-3138.

Dodd, PR. (2002). Excited to death: different ways to lose your neurones. *Biogerentology.* Vol.3, No.1-2, pp. 51-56.

Domico, LM., Zeevalk, GD., Bernard, LP., & Cooper, KR. (2006). Acute neurotoxic effects of mancozeb and maneb in mesencephalic neuronal cultures are associated with mitochondrial dysfunction. *Neurotoxicology.* Vol.27, No.5, (September 2006), pp. 816-825.

Ebadi, M., & Sharma, SK. (2003). Peroxynitrite and Mitochondrial Dysfunction in the Pathogenesis of Parkinson's Disease. *Antioxidant and Redox Signalling.*Vol.5, No.3, (June, 2003) pp. 319-335.

Elbaz, A., Dufouil, C., & Alpérovitch A. (2007). Interaction between genes and environment in neurodegenerative diseases. *Comptes Rendus Biologies.* Vol.330, No.4, (April 2007), pp. 318-328.

EPA. (2002). Registration Eligibility Decision for Endosulfan. EPA 738-R-02-013. July 2011, Available from: http://www.epa.gov/oppsrrd1/REDs/endosulfan_red.pdf

EPA. (2003). Revised Atrazine Interim Registration Eligibility Document. EPA-HQ-OPP-2003-0367-0003. August 2011, available from: http://www.epa.gov/oppsrrd1/REDs/atrazine_combined_docs.pdf

EPA. (2005). Registration Eligibility Decision (RED) for Maneb. EPA 738-R-05-XXX. August 2011, Available from: http://www.epa.gov/oppsrrd1/REDs/maneb_red.pdf

EPA. (2010). Endosulfan: final product cancellation order. Federal Register. Vol.75, No.217. FR Doc No: 2010-28138. EPA-HQ-OPP-2002-0262. July 2011, Available from: http://www.gpo.gov/fdsys/pkg/FR-2010-11-10/pdf/2010-28138.pdf

EPA. (2011). Pesticides Industry Sales and Usage 2006 and 2007 Market Estimates. September, 2011, Available from: http://www.epa.gov/opp00001/pesticides/market_estimate2007.pdf

EPA Persistent Bioaccumulative and Toxic (PBT) Chemical Program. (2011). August 2011, Available from: http://www.epa.gov/pbt/pubs/aldrin.htm

Fahn, S. (1999). Parkinson disease, the effect of levodopa, and the ELLDOPA trial. *Archives of Neurology*. Vol.56, (May, 1999) pp. 529-535.

Fahn, S., Oakes, D., Shoulson, Ira., Kieburtz, K., Rudolph, A., Lang, A., Olanow, W., Tanner, C., & Marek, K. (2004). Levodopa and Progression of Parkinson's disease. *New England Journal of Medicine*. Vol.351, (December, 2004) pp. 2498-2508.

Fei, Q., & Ethell, DW. (2008). Maneb potentiates paraquat neurotoxicity by inducing key Bcl-2 family members. *Journal of Neurochemistry*. Vol.105, No.6, (June 2008), pp. 2091-2097.

Fernagut, PO., Huston, CB., Fleming, SM., Tetreaut, NA., Salcedo, J., Masliah, E., & Chesselet, MF. (2007). Behavioral and histopathological consequences of paraquat intoxication in mice: Effects of α-synuclein over-expression. *Neuroscience*. Vol.61, No.12, (September, 2007) pp. 991-1001.

Ferrante, RJ., Schluz, JB., Kowall, NW., Beal, MF. (1997). Systemic administration of rotenone produces selective damage in the striatum and globus pallidus, but not the substantia nigra. *Brain Research*. Vol. 752, No.1, (April, 1997) pp. 157-162.

Filipov, NM., Stewart, MA., Carr, RL., & Sistrunk, SC. (2007). Dopaminergic toxicity of the herbicide atrazine in rat striatal slices. *Toxicology*. Vol.232, No.1-2, (March-April 2007), pp. 67-78.

Firestone, JA., Smith-Weller, T., Franklin, G., Swanson, P., Longstreth, WT., & Checkoway, H. (2005). Pesticides and risk of Parkinson's disease. A population-based case-control study. *Archives of Neurology*. Vol.62 No.1, (January 2005), pp. 91–95.

Franco, R., Li, S., Rodriguez-Rocha, H., Burns, M., & Panayiotidis, MI. (2010). Molecular mechanisms of pesticide-induced neurotoxicity: Relevance to Parkinson's disease. *Chemico-Biological Interactions*. Vol.188, No.2, (November 2010), pp. 289-300.

Gammon, DW., Aldous, CN., Carr, WC Jr., Sanborn, JR., & Pfeifer, KF. (2005). A risk assessment of atrazine use in California: human health and ecological aspects. *Pest Management Science*. Vol.61, No.4, (April 2005), pp. 331-355.

Gatto, NM., Cockburn, M., Bronstein, J., Manthripragada, AD., & Ritz, M. (2009). Well-water consumption and Parkinson's Disease in rural California. *Environmental Health Perspective*. Vol.117, No.12, (December, 2009) pp. 1912-1918.

Gauthier, E., Fortier, I., Couchesne, F., Pepin, P., Mortermer, J., & Gauvreau, D. (2001). Environmental pesticide exposure as a risk factor for Alzheimer's disease: a case-control study. *Environmental Research*. Vol.86, No.1, (May 2001), pp. 37-45.

Giusi, G., Facciolo, RM., Canonaco, M., Alleva, E., Belloni, V., Dessi-Fulgheri, F., & Santucci, D. (2006). The endocrine disruptor atrazine accounts for a dimorphic somatostatinergic neuronal expression pattern in mice. *Toxicological Sciences*. Vol.89, No.1, (January 2006), pp. 257-264.

Gorell, JM., Johnson, CC., Rybicki, BA., Peterson, EL., & Richardson, RJ. (1998). The risk of Parkinson's disease with exposure to pesticides, farming, well water, and rural living. *Neurology*. Vol.50, No.5, (May 1998), pp. 1346–1350.

Government of Canada (2009). Endosulfan: Canada's submission of information specified in Annex E of the Stockholm Convention pursuant to Article 8 of the Convention. July 2011, Available from:
http://chm.pops.int/Portals/0/docs/Responses_on_Annex_E_information_for_e ndosulfan/Canada_090110_SubmissionEndosulfanInformation.doc

Gysin, H., & Knuesli, E. (1960). Chemistry and herbicidal properties of triazine derivatives. In: *Advances in Pest Control Research*, R. Metcalf (Ed.), pp. 289-358, ISSN 0568-0107, Interscience, New York, USA.

Hatcher, JM., Richardson, JR., Guillot, TS., McCormack, AL., Di Monte, DA., Jones, DP., Pennell, KD., & Miller, GW. (2007). Dieldrin exposure induces oxidative damage in the mouse nigrostriatal dopamine system. *Experimental Neurology*. Vol.204, No.2, (April 2007), pp. 619-630.

Hatefi, Y. (1985). The Mitochondrial Electron Transport and Oxidative Phosphorylation System. *Annual Review of Biochemistry*. Vol.54, (July, 1985) pp. 1015-1069.

Hayden, KM., Northon, MC., Darcey, D., Ostbye, T., Zandi, PP., Breitner, JC., Welsh-Bohmer, KA.; & Cache County Study Investigators. (2010). Occupational exposure to pesticides increases the risk of incident AD: the Cache County Study. *Neurology*. Vol.74, No.19, (May 2010), pp. 1524-1530.

Hayes, WJ., & Laws, ER. (1991). *Handbook of pesticide toxicology*. Academic Press, ISBN 0123341604, San Diego, USA, pp. 816-822.

Hazardous Substances Databank. (1995). Paraquat. March, 1994. Available from: http://toxnet.nlm.nih.gov/cgi-bin/sis/search/f?./temp/~1ob7B1:1

Health Canada (2011). Discontinuation of Endosulfan. ISSN: 1925-0649; 1925-0630.

Heikkila, RE., Manzino, L., Cabbat, FS., Duvoisin, RC. (1984). Protection against the dopaminergic neurotoxicity of 1-methyl-4-phenyl-1,2,5,6-tetrahydropyridine by monoamine oxidase inhibitors. *Letters to Nature*. Vol. 311, (October, 1984) pp.467-469.

Hertzman, C., Wiens, M., Bowering, D., Snow, B., & Calne, D. (1990). Parkinson's disease: a case-control study of occupational and environmental risk factors. *American Journal of Industrial Medicine*. Vol.17, No.3, pp. 349–355.

Hertzman, C., Wiens, M., Snow, B., Kelly, S., & Calne, D. (1994). A case control study of Parkinson's disease in a horticultural region of British Columbia. *Movement Disorders*. Vol.9, No.1, (January 1994), pp. 69–75.

Hincal, F., Gürbay, A., & Giray, B. (1995). Induction of lipid peroxidation and alteration of glutathione redox status by endosulfan. *Biological Trace Element Research*. Vol.47, No.1-3, (January-March), pp. 321-326.

Hoglinger, GU., Lannuzel, A., Khondiker, ME, Michel, PP., Duyckaerts, C., Feger, J., Champy, P., Prigent, A., Medja, F., Lombes, A., Oertel, WH.,Ruberg, M., & Hirsch, EC. (2005). The mitochondrial cerebral inhibitor rotenone triggers a cerebral taupathy. *Journal of Neurochemistry*. Vol.95, No.4, (October, 2005) pp.930-939.

Hong, JS., Herr, DW., Hudson, PM., & Tilson, HA. (1986). Neurochemical effects of DDT in rat brain in vivo. *Archives of Toxicology. Supplement*. Vol.9, pp. 14-26.

Hossain, MM., & Filipov, NM. (2008). Alteration of dopamine uptake into rat striatal vesicles and synaptosomes caused by an *in vitro* exposure to atrazine and some of its metabolites. *Toxicology*. Vol.248, No.1, (June 2008), pp. 52-58.

Ikeda, T., Nagata, K., Shono, T., & Narahashi, T. (1998). Dieldrin and picrotoxinin modulation of GABA(A) receptor single channels. *Neuroreport*. Vol.9, No.14, (October 1998), pp. 3189-3195.

Inden, M., Kitamura, Y., Abe, M., Tamaki, A., Takata, K., & Taniguchi, T. (2010). Parkinsonian Rotenone Mouse Model: Re-evaluation of Long-Term Administration of Rotenone in C57BL/6 Mice. *Biological and Pharmaceutical Bulletin*. Vol.34, No.1, (October, 2010).

Jia, Z., & Misra, HP. (2007). Developmental exposure to pesticides zineb and/or endosulfan renders the nigrostriatal dopamine system more susceptible to these environmental chemicals later in life. *Neurotoxicology*. Vol.28, No.4, (July), pp. 727-735.

Kachroo, A., Irizarry, MC., & Schwarzschild, MA. (2010). Caffeine protects against combined paraquat and maneb-induced dopaminergic neuron degeneration. *Experimental Neurology*. Vol.223, No.2, (June 2010), pp. 657-661.

Kang, K., Park, J., Ryu, D., & Lee, Y. (2001). Effects and neuro-toxic mechanisms of 2, 2', 4, 4', 5, 5'-hexachlorobiphenyl and endosulfan in neuronal stem cells. Journal of Veterinary Medical Science. Vol.63, No.11, (November 1991), pp. 1183-1190.

Kanthasamy, AG., Kitazawa, M., Kanthasamy, A., & Anantharam, V. (2005). Dieldrin-induced neurotoxicity: relevance to Parkinson's disease pathogenesis. *Neurotoxicology*. Vol.26, No.4, (August 2005), pp. 701-719.

Karatas, AD., Aygun, D., & Baydin, A. (2006). Characteristics of endosulfan poisoning: a case study of 23 cases. *Singapore Medical Journal*. Vol.47, No.12, (December 2006), pp. 1030-1032.

Kiely, T., Donaldson, D., & Grube, A. (2004). Pesticides Industry Sales and Useage. 2000 and 2001 Market Estimates. US EPA. July 2011, Available from: http://www.epa.gov/opp00001/pestsales/01pestsales/ market_estimates2001.pdf

Kim, YH., Rane, A., Lussier, S., & Andersen, JK. (2011). Lithium protects against oxidative-stress-mediated cell death in a α-synuclein-overexpressing in vitro and in vivo models of Parkinson's disease. *Journal of Neuroscience Research*. Vol.89, No.10, (October 2011), pp. 1666-1675.

Kitazawa, M., Anantharam, V., & Kanthasamy, AG. (2001). Dieldrin-induced oxidative stress and neurochemical changes contribute to apoptopic cell death in dopaminergic cells. *Free Radical Biology and Medicine*. Vol.31, No.11, (December 2001), pp. 1473-1485.

Klaassen, CD., (Ed). (1996). *Casarett & Doull's toxicology: the basic science of poisons. 5th ed.* MacGraw-Hill Publishing Company. ISBN 0071470514, New York, USA, pp. 542-547.

Koller, W., Vetere-Overfield, B., Gray, C., Alexander, C., Chin, T., Dolezal, J, Hassanein, R., & Tanner, C. (1990). Environmental risk factors in Parkinson's disease. *Neurology.* Vol.40, No.8, (August 1990), pp. 1218–1221.

Kuopio, A-M., Marttila, RJ., Helenius, H., & Rinne, UK. (1999). Environmental risk factors in Parkinson's disease. Movement Disorders. Vol.14, No.6, (November 1999), pp. 928–939.

Lake, C., Wormstall, H., Einsiedler, K., & Buchkremer, G. (2004). Alzherimer's disease with secondary Parkinson's syndrome. Case report of a patient with dementia and Parkinson's syndrome after long-term occupational exposure to insecticides, herbicides, and pesticides. *Der Nervenarzt.* Vol.75, No.11, (November 2004), pp. 1107-1111.

Lakshmana, MK., & Raju, TR. (1994). Endosulfan induces small but significant changes in the levels of noradrenaline, dopamine and serotonin in the developing rat brain and deficits in the operant learning performance. *Toxicology.* Vol.91, No.2, (July 1994), pp. 139-50.

Landers, JE., Shi, L., Cho, TJ., Glass, JD., Shaw, CE., Leigh, PN., Diekstra, F., Polak, M., Rodriguez-Leyva, I., Niemann, S., Traynor, BJ., McKenna-Yasek, D., Sapp, PC., Al-Chalabi, A., Wills, AM., & Brown, RH Jr. (2008). A common haplotype within the PON1 promoter region is associated with sporadic ALS. *Amyotrophic Lateral Sclerosis.* Vo.9, No.5, (October 2008), pp. 306-314.

Langston, JW., Ballard, P., Tetrud, JW., & Irwin, I. (1993). Chronic Parkinsonism in humans due to a product of meperidine-analog synthesis. *Science.* Vol.219, No.4587, (February 1993) pp. 979–980.

Lee, E., Burnett, CA., Lalich, N., Cameron, LL., & Sestito, JP. (2002). Proportionate mortality of crop and livestock farmers in the United States, 1984–1993. *American Journal of Industrial Medicine.* Vol.42, No.5, (November 2002), pp. 410–20.

Lesnick, TG., Papapetropoulos, S., Mash, DC., Ffrench-Mullen, J., Shehadeh, L., Andrade, M., Henley, JR., Rocca, WA., Ahlskog, JE., & Maraganore, DM. (2007). A Genomic Pathway Approach to a Complex Disease: Axon Guidance and Parkinson Disease. *PLoS Genetics.* Vol.3, No.6, (June, 2007).

Li, X., Matsumoto, K., Murakami, Y., Tezuka, Y., Wu, Y., & Kadota, S. (2005). Neuroprotective effects of Polygonum multiflorum on nigrostriatal dopaminergic degeneration induced by paraquat and maneb in mice. *Pharmacology, Biochemistry and Behavior.* Vol.82, No.2, (October 2005), pp. 345-352.

Liou, HH., Tsai, MC., Chen, CJ., Jeng, JS., Chang, YC., Chen, SY., & Chen, RC. (1997). Environmental risk factors and Parkinson's disease: a case-control study in Taiwan. *Neurology.* Vol.48, No.6, (June 1997), pp. 1583–1588.

Manning-Bog, A.M., McCormack, AL., Li, J., Uversky, VN., Fink, AL., & Di Monte, DA. (2003). The Herbicide Paraquat Causes Up-Regulation and Aggregation of α-synuclein in Mice. *The Journal of Biological Chemistry.* Vol.277, (January, 2002) pp.1641-1644.

Matile, Ph., & Moor. (1967). Vacuolation: Origin and the development of the lysosomal apparatus in root tip cells. Planta. Vol.80, No.2, pp. 159-175.

McGuire, V., Longstreth, WT Jr., Nelson, LM., Koepsell, TD., Checkoway, H., Morgan, MS., & van Belle, G. (1997). Occupational exposures and amyotrophic lateral sclerosis. A population-based case-control study. American Journal of Epidemiology. Vol.145, No.12, (June 1997), pp. 1076-1088.

McNaught, KS., Olanow, CW., Halliwell, B., Isacson, O., & Jenner, P. (2001). Failure of the ubiquitin-proteasome system in Parkinson's disease. *Nature Reviews. Neuroscience.* Vol.2, No.8, (August 2001), pp. 589–594.

Michaelis, L., & Hill, ES. (1933). Potentiometric Studies on Semiquinones. *Journal of American Chemical Society.* Vol.55, No.4, (April 1933) pp. 1481-1494.

Moon, JM., & Chun, BJ. (2009). Acute endosulfan poisoning: a retrospective study. *Human and Experimental Toxicology.* Vol.28, No.5, (May 2009), pp. 309-316.

Morahan, JM., Pamphlett, R. (2006). Amyotrophic lateral sclerosis and exposure to environmental toxins: an Australian case-control study. *Neuroepidemiology.* Vol.27, No.3, pp. 130-135.

Occupational Health Services, Inc. (1991). MSDS for Maneb. OHS Inc., Secaucus, NJ, USA.

Obeso, JA., Olanow, CW., Nutt, JG. (2000). Levodopa motor complications in Parkinson's disease. *Trends in Neurscience.* Vol. 23, No. 1, (October, 2000) pp. S2-S7

Ossowska, K., Wardas, J., Smiatowska, M., Kuter, K., Lenda, T., Wieronska, JM., Zieba, B., Nowak, P., Dabrowska, J., Bortel, A., Kwiecinski, A., & Wolfarth, S. (2005). A slowly developing dysfunction of dopaminergic nigrostriatal neurons induced by long-term paraquat administration in rats: an animal model of preclinical stages of Parkinson's disease? *Neuroscience.* Vol.22, No.6, (September, 2005) pp. 1294-1304.

Pan, T., Rawal, P., Wu, Y., Xie, W., Jankovic, J., & Le, W. (2009). Rapamycin protects against rotenone-induced apoptosis through autophagy induction. *Neuroscience.* Vol.164, No.2, (December, 2009) pp. 541-551

Park, RM., Schulte, PA., Bowman, JD., Walker, JT., Bondy, SC., Yost, MG., Touchstone, JA., & Dosemeci, M. (2005). Potential occupational risks for neurodegenerative diseases. *American Journal of Industrial Medicine.* Vol.48, No.1, (July 2005), pp. 63-77.

Parrón, T., Requena, M., Hernández, AF., & Alarcón, R. (2011). Association between environmental exposure to pesticides and neurodegenerative diseases. *Toxicology and Applied Pharmacology.* (May 2011), Doi: 10.1016/j.taap.2011.05.006.

Paul, V., Balasubramaniam, E., & Kazi, M. (1994). The neurobehavioural toxicity of endosulfan in rats: a serotonergic involvement in learning impairment. *European Journal of Pharmacology.* Vol.270, No.1, (January 1994), pp. 1-7.

Peterson, RT., Nass, R., Boyd, WA., Freedman, JH., Dong, K., & Narahashi, T. (2008). Use of non-mammalian alternative models for neurotoxicological study. *Neurotoxicology.* Vol.29, No.3, (May 2008), pp. 546-555.

Pradhan, S., Pandey, N., Phadke, RV., Kaur, A., Sharma, K., & Gupta, RK. (1997). Selective involvement of basal ganglia and occipital cortex in a patient with acute endosulfan poisoning. *Journal of Neurological Sciences.* Vol.47, No.2, (April 1997), pp. 201-213.

Priyadarshi, A., Khuder, SA., Schaub, EA., & Priyadarshi, SS. (2001). Environmental risk factors and Parkinson's disease: a metaanalysis. *Environmental Research.* Vol.86, No.2, (June 2001), pp. 122-127.

Priyadarshi, A., Khuder, SA., Schaub, EA., Shrivastava, S. (2000). A meta-analysis of Parkinson's disease and exposure to pesticies. *Neurotoxicology.* Vol.21, No.4, (August 2000), pp. 435-440.

Quijano, RF. (2002). Endosulfan poisoning in Kasargod, Kerala, India. Report of a fact finding mission. Pesticide Action Network Asia and the Pacific. July 2011, Available at
http://www.panap.net/sites/default/ files/endosulfan_report_Kerala_1.pdf

Qureshi, MM., Hayden, D., Urbinelli, L., Ferrante, K., Newhall, K., Myers, D., Hilgenberg, S., Smart, R., Brown, RH., & Cudkowicz, ME. (2006). Analysis of factors that modify susceptibility and rate of progression in amyotrophic lateral sclerosis (ALS). *Amyotrophic Lateral Sclerosis.* Vol.7, No.3, (September 2006), pp. 173-182.

Ribaudo, MO., & Bouzaher, A. (1994). Atrazine: Environmental Characteristics and Economics of Management. United States Department of Agriculture. Agricultural Economic Report Number 699. July 2011, Available at http://www.ers.usda.gov/publications/AboutPDF.htm

Richardson, JR., Caudle, WM., Wang, M., Dean, ED., Pennell, KD., & Miller, GW. (2006). Developmental exposure to the pesticide dieldrin alters the dopamine system and increases neurotoxicity in an animal model of Parkinson's disease. *The FASEB Journal.* Vol.20, No.10, (August 2006), pp. 1695-1697.

Rocca, WA., Bonaiuto, S., Lippi, A., Luciani, P., Turtù, F., Cavarzeran, F., & Amaducci, L. (1990). Prevalence of clinically diagnosed Alzheimer's disease and other dementing disorders: A door-to-door survey in Appignano, Macerata Province, Italy. *Neurology.* Vol.40, No.4, (April 1990), pp. 626-663.

Rodríguez-Navarro, J., Rodríguez, L., Casarejos, MJ., Solano, RM., Gómez, A., Perucho, J., Cuervo, AM., de Yébenes, JG., & Mena, MA. (2010). Trehalose ameliorates dopaminergic and tau pathology in parkin deleted/tau overexpressing mice through autophagy activation. *Neurobiology of Diseases.* Vol.39, No. 3, (September, 2010) pp. 423-438.

Rodríguez, VM., Thiruchelvam, M., & Cory-Slechta, DA. (2005). Sustained exposure to the widely used herbicide atrazine: altered function and loss of neurons in brain monoamine systems. *Environmental Health Perspectives.* Vol.113, No.6, (June 2005), pp. 708-715.

Roede, J.R., Hansen, J.M., Go, Y.M., & Jones, D.P. (2011). Maneb and paraquat-mediated neurotoxicity: involvement of peroxiredoxin/thioredoxin system. *Toxicological Sciences.* Vol.121 No.2, (June), pp. 368-375.

Ross, MK., Jones, TL., & Filipov, NM. (2009). Disposition of the herbicide 2-chloro-4-(ethylamino)-6-(isopropylamino)-s-triazine (Atrazine) and its major metabolites in mice: a liquid chromatography/mass spectrometry analysis of urine, plasma, and tissue levels. *Drug Metabolism and Disposition: the Biological Fate of Chemicals.* Vol.37, No.4, (April 2009), pp. 776-786.

Saeed, M., Siddique, N., Hung, WY., Usacheva, E., Liu, E., Sufit, RL., Heller, SL., Haines, JL., Pericak-Vance, M., & Siddique, T. (2006). Paraoxonase cluster polymorphisms are associated with sporadic ALS. *Neurology.* Vol.67, No.5, (September 2006), pp. 771-776.

Sanchez-Ramos, J., Facca, A., Basit, A., & Song, S. (1998). Toxicity of dieldrin for dopaminergic neurons. *Experimental Neurology.* Vol.150, No.2, (April 1998), pp. 263-271.

Sano, M., Ernesto, C., Thomas, RG., Klauber, MR., Schafer, K., Grundman, M., Woodbury, P., Growdon, J., Cotman, CW., Pfeiffer, E., Schneider, LS., & Thal, LJ. (1997). A controlled trial of selegiline, alpha-tocopherol, or both as treatment for Alzheimer's disease. The Alzheimer's Disease Cooperative Study. *New England Journal of Medicine.* Vol.336, No.17, (April, 1997) pp. 1216-1222.

Seidler, A., Hellenbrand, W., Robra, B-P., Vieregge, P., Nischan, P., Joerg, J., Oertel, WH., Ulm, G., & Schneider, E. (1996). Possible environmental, occupational, and other etiologic factors for Parkinson's disease: a case-control study in Germany. *Neurology.* Vol.46, No.5, (May 1996), pp. 1275–1284.

Semchuk, KM., Love, EJ., & Lee, RG. (1992). Parkinson's disease and exposure to agricultural work and pesticide chemicals. *Neurology.* Vol.42, No.7, (July 1992), pp. 1328–1335.

Silva, MH., & Gammon, D. (2009). An assessment of the developmental, reproductive and neurotoxicity of endosulfan. *Birth Defects Research (Part B).* Vol.86, No.1, (February 2009), pp. 1-28.

Singh, ND., Sharma, AK., Dwivedi, P., Patil, RD., & Kumar, M. (2007). Citrinin and endosulfan induced maternal toxicity in pregnant Wistar rats: pathomorphological study. *Journal of Applied Toxicology.* Vol.27, No.6, (November-December), pp. 589-601.

Singhal, NK., Srivastava, G., Patel, DK., Jain, SK., & Singh, MP. (2011). Melatonin or silymarin reduces maneb- and paraquat-induced Parkinson's disease phenotype in the mouse. *Jounal of Pineal Research.* Vol.50, No.2, (March 2011), pp. 97-109.

Sirivarasai, J., Kaojarern, S., Yoovathaworn, K., & Sura, T. (2007). Paraoxonase (PON1) polymorphism and activity as the determinants of sensitivity to the organophosphates in human subjects. *Chemico-Biolical Interact.* Vol.168, No.3, (July), pp. 184-192.

Sivaraman, Valsala, Indulal, Gangadharan, Kalavathy, Beegom. (2003). Healh hazards of aerial spraying of endosulfan in Kasargod district, Kerala. Government of Kerala. July 2011, Available from:
http://endosulphanvictims.org/resources/KeralaGovt_FinalReport.pdf

Slowik, A., Tomik, B., Wolkow, PP., Partyka, D., Turaj, W., Malecki, MT., Pera, J., Dziedzic, T., Szczudlik, A., & Figlewicz, DA. (2006). Paraoxonase gene polymorphism and sporadic ALS. *Neurology.* Vol.67, No.5, (September 2006), pp. 766-770.

Smith, P., & Heath, D. (1976). Paraquat. *Critical Reviews in Toxicology.* Vol.4, No.4, (October, 1976) pp. 411-445.

Soderlund, DM., & Bloomquist, JR. (1989). Neurotoxic actions of pyrethroid insecticides. *Annual Review of Entomology.* Vol.34, pp. 77-96.

Somayajulu-Niţu, M., Sandhu, JK., Cohen, J., Sikorska, M., Sridhar, TS., Matei, A., Borowy – Borowski, H & Pandey, S. (2009). Paraquat induces oxidative stress, neuronal loss in substantia nigra region and Parkinsonism in adult rats: Neuroprotection and amelioration of symptoms by water-soluble formulation of Coenzyme Q10. *BMC Journal of Neuroscience.* Vol.10, No.88, (July 2009).

Sohal, RS., & Weindruch, R. (1996). Oxidative Stress, Caloric Restriction and Ageing. *Science.* Vol.273, No.5271, (July, 1995) pp. 59-63.

Stevens, JT., & Sumner, D D. (1991). Herbicides. In Handbook of Pesticide Toxicology. *Academic Press, New York.* Vol. 3, pp. 1317–1408.

Storch, A., Ludolph, AC., & Schwarz, J. (2004). Dopamine transporter: involvement in selective dopaminergic neurotoxicity and degeneration. *Journal of Neural Transmission.* Vol.111, No.10-11, (October 2004), pp. 1267-1286.

Sulkava, R., Heliovaara, M., Palo, J., Wikstrom, J., & Aromaa, A. (1988). Regional differences in the prevalence of Alzheimer's disease, In Soininen H (Ed): *Proceedings of the International Symposium on Alzheimer's disease.* ISBN 9517804881, Kuopio, Finland, University of Kuopio, June 12-15, 1988.

Takahashi, RN., Rogerio, R., & Zanin, M. (1989). Maneb enhances MPTP neurotoxicity in mice. *Research Communications in Chemical Pathology and Pharmacology.* Vol.66, No.1, (October 1989), pp. 167-70.

The Parkinson's Study Group. Effects of Tocopherol and Deprenyl on the Progression of Disability in Early Parkinson's Disease. (1993). *The New England Journal of Medicine.*Vol. 328. (January, 1994). Pp. 176-183.

Thiffault, C., Langston, WJ., Di Monte, DA. (2001). Acute exposure to organochlorine pesticides does not affect striatal dopamine in mice. *Neurotoxicology Research.* Vol.3, No.6, (November 2001), pp. 537-543.

Thiruchelvam, M., Brockel, BJ., Richfield, EK., Baggs, RB. & Cory-Slechta, DA. (2000a). Potentiated and preferential effects of combined paraquat and maneb on nigrostriatal dopamine systems: environmental risk factors for Parkinson's disease? *Brain Research.* Vol.873, No.2, (August 2000), pp. 225–234.

Thiruchelvam, M., Richfield, E.K., Baggs, R.B., Tank, A.W. & Cory-Slechta, DA. (2000b). The nigrostriatal dopaminergic system as a preferential target of repeated exposures to combined paraquat and maneb: implications for Parkinson's disease. *Journal of Neurosciences.* Vol.20, No.24, (December 2000), pp. 9207–9214.

Thiruchelvam, M., Richfield, EK., Goodman, BM., Baggs, RB., & Cory-Slechta, DA. (2002). Developmental exposure to the pesticides paraquat and maneb and the Parkinson's disease phenotype. *Neurotoxicology.* Vol.23, No.4-5, (October, 2002), pp. 321-333.

Thiruchelvam, M., Richfield, EK., Goodman, BM., Baggs, RB., & Cory-Slechta, DA. (2005). Developmental exposure to the pesticides paraquat and maneb and the Parkinson's disease phenotype. *Neurotoxicology.* Vol.23, No.4-5, (October 2005), pp. 621-633.

Tomita, M., Okuyama, T., Katsuyama, H., Miura, Y., Nishimura, Y., Hidaka, K., Otsuki, T., & Ikawa, TI. (2007). Mouse model or paraquat-poisoned lungs and its gene expression profile. *Toxicology.* Vol.231, No.2-3, (March, 2007) pp. 200-209.

Tsai, CH., Lo, SK., See, LC., Chen, HZ., Chen, RS., Weng, YH., Chang, FC., & Lu, CS. (2002). Environmental risk factors of young onset Parkinson's disease: a case-control study. *Clinical Neurology and Neurosurgery.* Vol.104, No.4, (September 2002), pp. 328-333.

Tuchsen, F., & Jensen, AA. (2000). Agricultural work and the risk of Parkinson's disease in Denmark, 1981–1993. *Scandinavian Journal of Work, Environment and Health.* Vol.26, No.4, (August 2000), pp. 359–362.

Uversky, VN. (2004). Neurotoxicant- induced animal models of Parkinson's disease: understanding the role of rotenone, maneb and paraquat in neurodegeneration. *Cell and Tissue Research.* Vol.318, No.1 (October, 2004) pp. 225-251.

Valdmanis PN, Kabashi E, Dyck A, Hince P, Lee J, Dion P, D'Amour M, Souchon F, Bouchard JP, Salachas F, Meininger V, Andersen PM, Camu W, Dupré N, Rouleau GA. (2008). Association of paraoxonase gene cluster polymorphisms with ALS in France, Quebec, and Sweden. Neurology. 71(7):514-520.

Wagner, SL. (1981). Clinical Toxicology of Agricultural Chemicals. *Environmental Health Science.* pp. 309.

Walsh, TJ., & Chrobak, JJ. (1987). The use of the radial arm maze in neurotoxicology. *Physiology and Behavior.* Vol.40, No.6, pp. 799-803.

Weed Science Society of America, 1994.

Weisskopf, MG., Morozova, N., O'Reilly, EJ., McCullough, ML., Calle, EE., Thun, MJ., & Ascherio, A. (2009). Prospective study of chemical exposures and amyotrophic lateral sclerosis. *Journal of Neurology, Neurosurgery and Psychiatry*. Vol.80, No.5, (May 2009), pp. 558-561.

Weisskopf, MG., Knekt, P., O'Reilly, EJ., Lyytinen, J., Reunanen, A., Kaden, F., Altshul, L., & Ascherio, A. (2010). Persistent organochlorine pesticides in serum and risk of Parkinson disease. *Neurology*. Vol.74, No.13, (mo), pp. 1055-1061.

Wesseling, C., Joode, WD., Rupert, C., Leon, C., Monge, P., Hermosillo, H., & Partnen, T J. (2001). Paraquat in Developing Countries. *International Journal of Occupational Environmental Health*. Vol.7, No.4, (October, 2001) pp. 275–286

Wills, AM., Cronin, S., Slowik, A., Kasperaviciute, D., Van Es, MA., Morahan, JM., Valdmanis, PN., Meininger, V., Melki, J., Shaw, CE., Rouleau, GA., Fisher, EM., Shaw, PJ., Morrison, KE., Pamphlett, R., Van den Berg, LH., Figlewicz, DA., Andersen, PM., Al-Chalabi, A., Hardiman, O., Purcell, S., Landers, JE., & Brown, RH Jr. (2009). A large-scale international meta-analysis of paraoxonase gene polymorphisms in sporadic ALS. *Neurology*. Vol.73, No.1, (July 2009), pp. 16-24.

Wirdefeldt, K., Adami, HO., Cole, P., Trichopoulos, D., & Mandel, J. (2011). Epidemiology and etiology of Parkinson's disease: a review of the evidence. *European Journal of Epidemiol*. Suppl 1 (June 2011), pp. S1-58.

www2.syngenta.com/en/about_syngenta/companyhistory.html

www.zymes.com

Xu, HY., Chen, RR., Cai, XY., & He, DF. (2011). Effects of co-exposure to paraquat and maneb on system of substantia nigra and striatum in rats. *Zhonghua Lao Dong Wei Sheng Zhi Ye Bing Za Zhi*. Vol.29, No1, (January 2011), pp. 33-38.

Yang, X., Jiang, Y., Weng, J., Wang, J. (2004). Study on the influence on the proliferation and the induction of apoptosis on PC 12 cells by endosulfan. *Chinese Journal of Clinical Neurosciences*. 2004-1.

Zaidi, NF., Agrawal, AK., Anand, M., & Seth, PK. (1985). Neonatal endosulfan neurotoxicity: behavioral and biochemical changes in rat pups. *Neurobehavioral Toxicology and Teratology*. Vol.7, No.5, (September-October 1985), pp. 439-442.

Zhang, J., Fitsanakis, VA., Gu, G., Jing, D., Ao, M., Amarnath, V., & Montine, TJ. (2003). Manganese ethylene-bis-dithiocarbamate and selective dopaminergic neurodegeneration in rat: a link through mitochondrial dysfunction. *Journal of Neurochemistry*. Vol.84, No.2, (January 2003), pp. 336-346.

Zhou, Y., Shie, FS., Piccardo, P., Montine, TJ., & Zhang, J. (2004). Proteasomal inhibition induced by manganese ethylene-bis-dithiocarbamate: relevance to Parkinson's disease. *Neuroscience*. Vol.128, No.2, pp. 281-91.

9

Herbicides Persistence in Rice Paddy Water in Southern Brazil

Renato Zanella[1], Martha B. Adaime[1], Sandra C. Peixoto[1],
Caroline do A. Friggi[1], Osmar D. Prestes[1], Sérgio L.O. Machado[1],
Enio Marchesan[1], Luis A. Avila[2] and Ednei G. Primel[3]
[1]Federal University of Santa Maria, Santa Maria - RS,
[2]Federal University of Pelotas, Pelotas - RS,
[3]Federal University of Rio Grande, Rio Grande - RS,
Brazil

1. Introduction

Although the agricultural activity is only one of the several sources of water pollution, it is thought to be an important cause on reducing the water quality through pollution by agrochemicals, in special pesticides. Environmental water pollution by pesticides is a topic of current international concern with widespread ecological consequences. The selective use of pesticides to control pests (insects, weeds and diseases) and vectors of plant diseases can aims at the increase of production of food crops. In agricultural production, herbicides are often used to efficiently control of weeds, however there is widespread concern over the effects of these synthetic chemicals on native fauna and flora. The risks of herbicides use to kill or otherwise manage certain species of plants considered to be pests need to be balanced against the benefits to the production. Plant pests, or weeds, compete with desired crop plants for light, water, nutrients and space. To reduce the intensity of the negative effects of weeds on the productivity of desired agricultural crops, fields may be sprayed with an herbicide that is toxic to the weeds, but not to the crop species. Consequently, the pest plants are selectively eliminated, while maintaining the growth of the desired plant species.

The degradation of applied pesticides or their conversion into other products does not necessarily mean the loss of biological activity, and many times, this conversion can result in even more toxic products. The study of the persistence of pesticides in crop fields is of great importance in order to evaluate the risks of environmental pollution. Due to the model of agriculture adopted in Brazil, pesticide usage has become intensive and many related environmental problems are occurring (Steckert et al., 2009). The characteristics of rice fields, the climate conditions and the use of pesticides contribute to the enhanced risk of surface water pollution, justifying the need to quantify their degree of occurrence and to implement measures to prevent it.

The irrigated agriculture of the Rio Grande do Sul State, in the Southern Brazil, is responsible for more than 60% of the Brazilian rice production (CONAB, 2010), corresponding to more than one million hectares. The cultivation of irrigated rice can generates a great impact to the environment, so much in amount as in the quality of the surrounding water because demands intense agrochemical use, mainly herbicides, insecticides and fertilizers (Noldin et al., 2001).

In Brazilian conditions, the contribution of agriculture to water pollution is not well quantified, but in United States, 50 to 61% of the pollution load that affects lakes and rivers come from agriculture (Gburek & Sharpley, 1997) being the superficial runoff the main mechanism of pesticides transport. Although agriculture is just one of the countless nonpoint-sources of pollution, it is generally targeted as the largest source among all pollutant categories. In Brazil, the culture of irrigated rice is notable in the Rio Grande do Sul State, with more than 70% of the Brazilian cultivated area of this culture, using circa of 5000 m^3 of water per ha (Machado et al., 2006).

In most of the rice farms, the pesticide applications is followed by the irrigation and depending on the handling of the water and on the occurrence of rain after pesticide application, there is a risk that part of the applied compounds will be carried out of the area, contaminating water sources (Jury et al., 1990; Squillace & Thurman, 1992; Solomon et al. 1996; Primel et al., 2005).

Irrigated farming for food production is the agricultural practice that most contributes to the deviation of water from its natural courses. Scarcity of water resources, which is occurring on a world-wide level, as well as the use of a large quantity of water, which is partially returned to its natural sources, makes irrigated rice farming a serious concern in terms of possible consequences for the environment, both in quantity and quality of the water sources (Kurz et al., 2009).

Rice productivity levels in Brazil are high, reaching an average value of approximately 6.0 t ha^{-1}, similar to that obtained in countries with tradition in irrigated rice farming, such as the United States, Australia and Japan. However, this high productivity is associated with the intense use of pesticides (Baird & Cann, 2005). Herbicides are potential contaminants of environmental water because they are directly applied to the soil or irrigating water. Thus, they can be leached to the surface water and transported into the groundwater (Hatrík & Tekel, 1996; Zanella et al., 2002). According to specialized literature (Barceló & Hennion, 1997; Primel et al.; 2005; Cabrera et al., 2008), a pesticide can pollute the aquatic environmental if its solubility in water is higher than 30 mg L^{-1}; its K_{oc}, organic carbon partition coefficient, is less than 300-500; its K_H, Henry Law constant is less than 10^{-2} Pa m^3 mol^{-1}, its soil half-life is longer than 2-3 weeks and its water half-life is longer than 25 weeks.

In the pre-germinated system of irrigated rice cultivation frequently employed in Brazil, drainage from the area after sowing can set off serious environmental problems, as well as cause the loss of nutrients and/or pesticides that are in suspension in the irrigation water that is released. This has been evidenced in studies carried out by Primel (2003) and Machado (2003), where the occurrence of some herbicides, mainly those that present high persistence, was confirmed in river and irrigation waters. In Brazil some studies has been done until now to investigate the behavior and destination of herbicides in the systems related to the rice paddy fields in an attempt to evaluate the risk to the environment. (Zanella et al. 2002; Noldin et al., 1997: Bortoluzzi et al. 2006; Marchesan et al. 2007; Resgalla et al., 2007; Bortoluzzi et al. 2007; Grützmacher et al. 2008; Caldas et al. 2009; Silva et al., 2009; Caldas et al. 2010; Demoliner et al. 2010; Marchesan et al. 2010; Reimche, 2010).

Herbicides can became a non-point source of pollution and, once in water, they spread in the environment, being hard to avoid their dispersal and their action on non-target organisms. Persistent and high-mobility herbicides have been detected in surface waters (Pereira & Rostad, 1990; Thurman et al. 1991; Thurman et al. 1992; Huber et al. 2000) and underground (Walls et al., 1996; Kolpin et al., 1998), meaning risks for the environment, especially for the quality of water. In rice paddy fields, flooding increases the chance of herbicide transport through rain water or tailwater runoff, increasing the potential of environmental pollution. Due to cultural practices that include soil tillage, intensive use of irrigation by flooding, and high intensity of fertilizers and pesticides application, the rice paddy presents potential risk of decrease water quality. This problem can be even more intense in pre-germinated rice system where the initial drainage after soil preparation not only results in losses of considerable amounts of water but also carry sediments and, consequently, the loss of nutrients and pesticides adsorbed to the sediments or dissolved in the solution. Moreover, in most rice paddy, the applications of herbicides are followed by flooding and in some cases, the pesticides, especially some herbicides and insecticides, are applied directly in irrigation water. According to the water management adopted and to the rainfall after the application, there is a risk that the residues of these products are carried outside of the area and contaminating the water bodies, even though the concentration of herbicides in irrigation water is, in general, low (Squillace & Thurmann, 1992; Sudo et al. 2002; Añasco et al., 2010).

Several surface water monitoring programs have been carried out to quantify the degree of water pollution by pesticides related to rice paddy fields (Kammerbauer & Moncada, 1998; Huber et al., 2000; Kolpin et al., 2000; Bouman et al., 2002; Carabias-Martínez et al., 2002; Laganà et al., 2002; Cerejeira et al., 2003; Carabias-Martínez et al., 2003; Palma et al., 2004; Konstantinou et al., 2006; Marchesan et al., 2007; Siemering et al., 2008; Baugros et al., 2008; Woudneh et al., 2009; Añasco et al., 2010). Analysis of temporal variations of pesticides shows that herbicides present relatively higher concentrations in the earlier stages of the rice planting season, while insecticides and fungicides have relatively higher concentrations at the later stages (Añasco et al., 2010).

Rice crop conducted under flooded conditions is pointed out as being an activity of high pollution potential. The factors that contribute to this claim are the large amount of water used to maintain the flood, the usual proximity of the fields to surface water bodies, the predominant shallow aquifer in these areas, and the intentional and unintentional release of water from the field (Machado et al., 2006). Several studies investigated the extension of herbicides runoff from paddy fields to the environment. In general, high herbicide concentrations were usually observed in the stream water following pesticide application periods and are generally detected in concentrations in levels of ng L^{-1} for only 2 to 3 months after use. (Ueji & Inao, 2001; Nakano et al., 2004; Numabe & Nagahora, 2006; Son et al., 2006; Comoretto et al., 2008).

2. Objective

The aim of this chapter is to present results from the evaluation of the persistence of the currently used herbicides bentazon, bispyribac sodium, clomazone, imazapic, imazethapyr, metsulfuron-methyl, penoxsulam, propanil, quinclorac and 2,4-D in rice paddy water from fields located in the Rio Grande do Sul State, in Southern Brazil, in order to assess the pollution risk of water from surrounding areas. Field experiments were conducted in irrigated rice experiments carried out in the Weed Science Department of the Federal University of Santa Maria (UFSM). Analyses were performed in the Laboratory of Analysis of Pesticide Residues (LARP) from the Federal University of Santa Maria (UFSM). Water samples were collected periodically and the herbicide residues were preconcentrated by solid phase extraction (SPE) followed by the determination by Liquid Chromatography with Diode Array Detection (LC-DAD) or Liquid Chromatography coupled to tandem Mass Spectrometry (LC-MS/MS).

3. Methods

3.1 Characterization of the physico-chemical properties of the herbicides

The more important properties of the studied herbicides are described in this section. Table 1 presents information about the evaluated herbicides.

Bentazon, 3-(1-methylethyl)-(1H)-2,1,3-benzothiadiazin-4(3H)-one 2,2-dioxide, is a post-emergence benzothiadiazinol herbicide used for selective control of broadleaf weeds and sedges in crops like rice, corn, beans and peanuts. Its selectivity is based on the ability of the crop plants to metabolize bentazon quickly to 6-OH- and 8-OH-bentazon and conjugate these with sugars, while weeds do not, so that photosynthesis is disrupted and the weeds die (Huber & Otto, 1994). The small K_{ow} value precludes bioaccumulation. Technical and formulated forms of bentazon are classified by EPA as practically nontoxic to fish. Bentazon does not bind to, or adsorb, to soil particles and it is highly soluble in water. These characteristics usually suggest a strong potential for groundwater contamination. Its rapid degradation is expected to prevent the contamination of groundwater. EPA estimates that bentazon may be found in about 0.1% of the rural drinking water wells nationwide. Bentazon has the potential to contaminate surface water because of both its mobility in runoff water and its pattern of use on rice. Bentazon appears to be stable to hydrolysis and has a half-life of less than 24 h in water because it undergoes photodegradation (EPA, 1985).

Bispyribac-sodium, sodium 2,6-bis(4,6-dimethoxypyrimidin-2-yloxy)benzoate, is a herbicide registered for use in irrigated rice farming in Brazil and in several other countries. It is indicated for the post-emergence control of grasses. This herbicide belongs to the toxicological class II, considered very toxic. The compound has a half-life in soil of less than 10 days (Ware, 1994; Vencill, 2002). Bispyribac-sodium presents a lethal dose of 3524 mg kg^{-1} when applied orally in rats, and presents CL_{50} for trout (96 h) > 100 mg L^{-1} and CL_{50} for Daphnia (48 h) > 100 mg L^{-1} (EPA, 2001; Tomlin, 2007).

By the Goss classification, considering the physicochemical properties, bispyribac sodium presented intermediate probabilities to be found in surface waters, showing Medium Water-Phase-Transport Runoff Potential (MWTRP) and Medium Sediment-Transport Runoff Potential (MSTRP) (Caldas et al., 2011).

Clomazone, 2-[(2-chlorophenyl)methyl]-4,4-dimethyl-3-isoxazolidinone, is moderately mobile in sandy soils and its half-life in soils range from 5 to 117 days, depending of the soil and climatic conditions (Curran et al., 1992; Kirksey et al., 1996; Mervosh et al., 1995). Studies show that the clomazone concentration in soil solution is dependent on the amount of carbon and water in the soil (Lee et al., 2004). The clomazone is widely used against species of annual broad leaf weeds and Grass in the cultivation of soybeans, cotton, rice, sugar cane, corn, tobacco and a variety of other vegetable crops (Zanella et al., 2000). It is stable at room temperatures for at least 2 years and it is also stable at 50 °C for at least 3 months. The properties of clomazone indicates its potential for aquatic environmental pollution and has been detected in the majority of the water samples collected from rivers located close to irrigated rice fields in the South of Brazil (Zanella et al., 2002).

Imazethapyr, 5-ethyl-2-[(RS)-4-isopropyl-4-methyl-5-oxo-2-imidazolin-2-yl]nicotinic acid, and **imazapic**, 2-[(RS)-4-isopropyl-4-methyl-5-oxo-2-imidazolin-2-yl]-5-methylnicotinic acid, are imidazolinones herbicides sold as a commercial mixture containing 75 and 25 g active ingredient L^{-1} of imazethapyr and imazapic, respectively. This product is applied in pre- and post-emergence for the control of red rice which cause damages in the production and commercialization of the grains (Kraemer, 2009).

The persistence of these herbicides in soil is influenced by the pH (Loux & Reese, 1992), by the humidity (Baughman & Shaw, 1996) and organic matter (Stougaard et al., 1990; Alister & Kogan, 2005). The main dissipation mechanisms are the microbiological degradation (Goetz et al., 1990) and the photolytic decomposition (Mallipudi et al., 1991). Imazethapyr and imazapic undergoes limited biodegradation in anaerobic conditions (Senseman, 2007; Santos et al. 2008).

Metsulfuron-methyl, methyl 2-(4-methoxy-6-methyl-1,3,5-triazin-2-ylcarbamoylsulfamoyl) benzoate, is a residual sulfonylurea compound widely used as a selective pre- and post-emergence herbicide due to its selectivity against a wide range of weeds in cereal, pasture, and plantation crops. It is a systemic compound with foliar and soil activity, and it works rapidly after it is taken up by the plant. Metsulfuron methyl is expected to have moderate to very high mobility and if released into water is expected to have little to no adsorption to suspended solids and sediment. Metsulfuron methyl is expected to biodegrade in water based on its behavior in soil.

Penoxsulam, 2-(2,2-difluoroethoxy)-N-(5,8-dimethoxy[1,2,4]triazolo[1,5-c]pyrimidin-2-yl)-6-(trifluoromethyl)benzenesulfonamide, is a sulfonamide herbicide for post-emergence control of grasses, sedges and broadleaf weeds in paddy rice. Soil sorption values are inversely related with the pH, indicating that penoxsulam is qualitatively mobile. EPA (2004) has classified this herbicide as a reduced risk pesticide but has also concluded that there are uncharacterized risks, particularly to plants and microbial communities.

Propanil, 3',4'-dichloropropionanilide, is one of the most used herbicides in the cultivation of irrigated rice in Brazil. It is a post-emergent selective herbicide of contact with short duration used to control the electron transport inhibition of photosynthesis in herbs of wide leaves (Tomlin, 2007). According to Barceló et al. (1998) it was also one of the most used in Tarragosa (Spain) and according to Coupe et al. (1998) it was extensively used in the area of the Mississipi Delta (USA). Various studies have pointed out that propanil is degraded quickly into 3,4-dichloroaniline (3,4-DCA) (Dahchour et al., 1986; Correa & Steen, 1995; Barceló & Hennion, 1997). Propanil is weakly adsorbed by the soil, is moderately mobile in sandy soils and of low mobility in clayey soils, with half-life of 1 to 3 days (Vencill, 2002).

Quinclorac, 3,7-dichloroquinoline-8-carboxylic acid, is widely used in post-emergence against species broad leaf weeds in the cultivation of irrigated rice. Quinclorac has variable mobility depending on soil type and organic matter and it can persist in the soil for one year affecting susceptible crops in rotation programs (Vencill, 2002). Quinclorac presented intermediate probabilities to be found in surface waters, showing Medium Water-Phase-Transport Runoff Potential (MWTRP) (Caldas et al., 2011).

2,4-D, (2,4-dichlorophenoxy)acetic acid, is a selective herbicide used to kill broadleaf weeds. 2,4-D is one of the most widely used herbicides in industrial, commercial, and government markets. Crops treated with 2,4-D include rice, corn, soybeans, wheat, sugarcane, and barley. Introduced in 1946, 2,4-D continues to be one of the most important herbicides across the globe. 2,4-D is readily broken down by microbes in soil and aquatic environments; half-life in soil is 7-10 days and the half-life in water is 3-28 days. There is little tendency of 2,4-D to bioconcentrate. Leaching to groundwater may occur in coarse and sandy soil that has a low organic content. Agricultural run-off containing 2,4-D may contaminate groundwater, which may impact drinking water in some areas (EPA, 2005). Because 2,4-D has demonstrated toxic effects on the thyroid and gonads following exposure, there is concern over potential endocrine-disrupting effects.

The herbicide 2,4-D has a low binding affinity in mineral soils and sediment, and in those conditions is considered intermediately to highly mobile. Although 2,4-D is highly mobile, rapid mineralization rates may reduce the potential of 2,4-D to affect groundwater (Boivin et al., 2005). The compound 2,4-D has been detected in streams and shallow groundwater at low concentrations, in both rural and urban areas (McPherson et al., 2003). In a monitoring study, traces of the herbicide 2,4-D were detected in 49.3% of finished drinking water samples and 53.7% of untreated water samples, with detections between 0.001 and 2.4 μg L^{-1} (USDA, 2007).

According to the GOSS criteria, regarding the water-phase-transport runoff potential, the herbicides imazethapyr and metsulfuron-methyl show high potential. Clomazone and imazapic besides presenting High Water-Phase-Transport Runoff Potential (HWTRP) showed Low Sediment-Transport Runoff Potential (LSTRP), which is a high indicative that these compounds have good chance to be found in surface waters. Some compounds such as penoxsulam show Low Water-Phase-Transport Runoff Potential (LWTRP) showing fewer tendencies to be found in surface waters. Bispyribac sodium and quinclorac due to their physicochemical properties, presented intermediate probabilities to be found in surface waters, in other words, Medium Water-Phase-Transport Runoff Potential (MWTRP) or associated with sediment (MSTRP). For the risk of contamination of the groundwater, according to the methods GUS and US-EPA, the herbicides bentazone, clomazone, imazethapyr, imazapic, metsulfuron-methyl, and quinclorac are classified as potential contaminants of groundwater. The herbicide bispyribac-sodium did not show any leaching tendency by both methods. Penoxsulam, propanil and 2,4-D show different classifications by the US-EPA and GUS methods, or due to the lack of physicochemical parameters the prediction was not possible. These differences between the methods can occur because they consider different physicochemical characteristics (Caldas et al., 2011).

The main physicochemical properties of the herbicides related to the persistence in water are: solubility in water, vapor pressure (VP), octanol/water partition coefficient (K_{ow}), acid dissociation constant (pK_a), Henry's Law Constant (H) and half-life ($t_{1/2}$). These properties can be used to estimate the risk of environmental pollution (Primel et al., 2005). According to Barceló & Hennion (1997), acid pesticides presenting pK_a < 3-4, basic pesticides: pK_a > 10;

polar have log K_{ow} < 1-1.5, non-polar: log K_{ow} > 4-5, and between these values the compounds are considered moderate polar. The pesticides with log K_{ow} >3.0 undergoes bioaccumulation. The K_{oc} is the soil organic carbon / water partition coefficient, which is the ratio of the mass of a chemical that is adsorbed in the soil per unit mass of organic carbon in the soil per the equilibrium chemical concentration in solution. K_{oc} values depend on the hydrophobicity of the compounds and are useful in predicting the mobility of organic contaminants in soil. Higher K_{oc} values correlate to less mobile organic compounds while lower K_{oc} values correlate to more mobile organic contaminants.

Herbicides (CAS number)	Chemical structure	Solubility in water, mg L^{-1}	K_{oc}, $cm^3 g^{-1}$	Log K_{ow}	pK_a	VP, mPa (20°C)	KH, Pa $m^3 mol^{-1}$
Bentazon (25057-89-0)		570	34	5.8	3.2	0.17	7.4×10^{-5}
Bispyribac sodium (125401-92-5)		73300	5000	-1.03	3.35	5.5×10^{-6}	3.1×10^{-11}
Clomazone (81777-89-1)		1100	150-562	2.54		19.2	4.2×10^{-3}
Imazapic (81334-60-3)		2200	137	2.47	2.0; 3.6	<0.013	n.a.
Imazethapyr (81335-77-5)		1400	10	1.49	2.1; 3.9	<0.013	1.3×10^{-2}
Metsulfuron-methyl (74223-64-6)		2790	2.9-27	-1.7	3.75	1.1×10^{-7}	4.5×10^{-11}
Penoxsulam (219714-96-2)		408	73.2	-0.60	5.1	2.5×10^{-11}	2.9×10^{-14}
Propanil (709-98-8)		130	239-800	3.3		0.026	3.6×10^{-3}
Quinclorac (84087-01-4)		0.065	50	-1.15	4.3	< 0.01	$<3.7 \times 10^{-2}$
2,4-D (94-75-7)		311	60	2.6-2.8	2.73	0.019	1.3×10^{-5}

Table 1. Chemical structure and some physicochemical properties of the selected herbicides. K_{oc} = soil/water partition coefficient; K_{ow} = octanol/water partition coefficient, pK_a= acid dissociation constant; VP= vapor pressure; KH = Henry's Law constant. (Barceló & Hennion, 1997; Tomlin, 2007; Senseman, 2007; Dores & De-Lamonica-Freire, 2001).

3.2 Field experiment

The experimental fields used in the studies were from the Weed Science Department at the Campus of the Federal University of Santa Maria (UFSM), Rio Grande do Sul (Brazil), and the experiments were conducted in different rice planting season (1999 to 2010). Figure 1 present the where the experiments were conducted. The fields have a typical Albaqualfa soil, with medium texture, presenting the following characteristics: 44% sand, 32% silt, 23% clay and 1.8% organic material. Soil was prepared with disking and graded by water slide. A randomized complete block design with four replications was used, with plots of 4 x 4 m. The irrigation of each plot was done individually through a water-pump linked to a float hydrometer to keep a 10-cm deep water paddy. The treatments were with the application of the commercial formulated herbicides at recommended dose, in g active ingredient ha^{-1}, how presented in Table 2. The herbicides were applied using CO_2-pressurized backpack sprayers. Taking into account a 10-cm water slide, the theoretical initial concentrations, in μg L^{-1}, estimated for each herbicide is informed in Table 2.

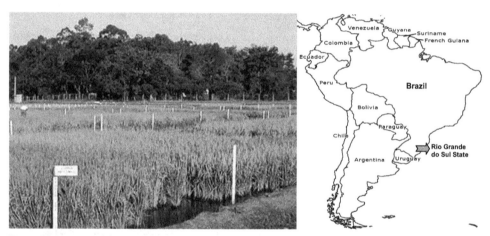

Fig. 1. The rice paddy field where the experiments were conducted. Location: latitude 29° 48´ 8.2″ and longitude 53° 43´ 22.6″

3.3 Determination of herbicide residues in water samples by Liquid Chromatography

Analytical methods were developed and validated for the determination of the studied herbicide residues in paddy water using Liquid Chromatography with Diode Array Detection (LC-DAD) or Liquid Chromatography coupled with tandem Mass Spectrometry (LC-MS/MS). Samples were analyzed according the methods developed by Primel (2003), Zanella et al. (2003); Caldas et al. (2010) and Demoliner et al. (2010).

Aliquots of samples, previously filtered in membranes of 47 mm of diameter and 0.45 µm of porosity were pre-concentrated in solid phase extraction (SPE) cartridges. The herbicides were eluted from the cartridges and the concentrations of the compounds were determined by LC-DAD or by LC-MS/MS with electrospray ionization (ESI). In order to achieve similar method limit of detection (LODm) values, the samples analyzed by LC-DAD were submitted to a higher preconcentration rate than the samples analyzed by LC-MS/MS. The LODm values for the selected herbicides are presented in Table 2.

The parameters of validation of the method include analytical curve, linearity, limit of detection (LOD), limit of quantification (LOQ), precision, in terms of repetitivity and intermediate precision, and accuracy (recovery). Paddy water samples free of herbicides were used to assess the extraction efficiency of the proposed method and to observe the interfering peaks.

3.3.1 Chemicals

Herbicide standards were purchased from Dr. Ehrenstorfer GmbH (Augsburg, Germany). The stock solutions were prepared in methanol at a concentration of 1000 mg L^{-1} and kept at -16 °C. The analytical solutions were prepared by dilution of this stock solution in the mobile phase. Solvents were Nanograde® degree (Mallinckrodt, USA), water was purified in the system Direct-Q UV3® (resistivity 18.2 MΩ cm, Millipore, USA).

3.3.2 Sampling

Water samples were collected directly in 1 L amber glass bottles in each sampling site at different days after treatment (DAT). These bottles had been cleaned prior to sampling and was filled with the water sample to the top with as little remaining air as possible, and sealed tightly. All samples were properly labeled with details of the source and sampling date, and stored at 4 °C until the preconcentration step, which was carried out on the same day of the sampling.

3.3.3 Liquid chromatography analyzes

LC-DAD. Liquid chromatograph system from Varian (Palo Alto, USA) with a solvent delivery system model 9002, diode array detector (DAD) ProStar 335, system of data acquisition Star Workstation 6.0, analytical column Synergi 4 μ Fusion RP-80 (250 x 4.6 mm i.d.; 4 μm) and guard-column of the same material (4 x 3 mm), both from Phenomenex (Torrance, USA) was used.

LC-MS/MS. Analyses of the investigated herbicides were performed on a Waters Alliance 2695 Module HPLC, equipped with a quaternary pump, an automatic injector and a thermostatted column compartment, and detection on a Quattro micro API (triple quadrupole) mass spectrometer, equipped with an electrospray (ESI) ionization source (Waters, Milford, MA, USA). The chromatographic separation was performed with an XTerra® MS C18 (3.0 mm × 50 mm i.d., 3.5 μm) column Waters (Milford, MA, Ireland). Analytical instrument control, data acquisition and treatment were performed by the software MassLynx, version 4.1 (Micromass, Manchester, UK).

The main advantage of the LC-MS/MS method is the use of MS/MS detector because it provides a high sensitivity and permits the confirmation of the herbicides identity.

3.3.4 Sample preparation

The samples were acidified to pH 3.0 with phosphoric acid prior the preconcentration step by SPE in cartridges containing octadecylsilane (C_{18}). The cartridges were conditioned with 3 mL of methanol, 3 mL of ultrapure water and 3 mL of ultrapure water acidified at pH 3.0. After the conditioning step, water samples were percolated through the cartridges using a SPE manifold.

For the LC-DAD analysis, samples were submitted to a high preconcentration rate to achieve the necessary method LOD. When LC-MS/MS was used, the sample preparation step was simplified using a low preconcentration rate, resulting in a faster method to perform. The use of LC-MS/MS allows the quantification with confirmation of the identity of herbicides.

3.4 Determination of the half-life in paddy water

The more stable the compounds, the longer it takes to break down. This can be measured in terms of its half life ($t_{1/2}$), the time taken for the concentration of the compounds to be reduced to 50% of the initial value. The longer it takes to break down, the higher its persistence. The half life is unique to individual products but variable depending on specific environmental and application conditions.

Herbicides half-life values in water were calculated according to the following equation:

$$t_{1/2} = \ln 2 \ / \ k = 0.693 \ / \ k \tag{1}$$

where $t_{1/2}$ is the herbicide half-life in days, ln is the natural logarithm and k is the constant of the rate of the herbicides dissipation in water (Barceló & Hennion, 1997). The calculation of the degradation constant (k) was performed using the first-order rate equation:

$$-\ln [C_t]/[C_o] = k \ . \ t \tag{2}$$

where C_t represents the concentration at time t; C_o represents the initial concentration; and k is the degradation constant, obtained by the inclination of the straight line.

4. Results and discussion

4.1 Sample preparation and determination of herbicides residues in water samples

The liquid chromatographic methods by LC-DAD and LC-MS/MS proved to be a good option for the determination of herbicide residues in paddy water, allowing the analysis with high selectivity and sensitivity. The method LOD values for herbicide residues in paddy water presented in Table 2, considering the pre-concentration step, were in the range of 0.05 to 0.2 µg L^{-1}. Good results for the preconcentration of herbicides from paddy water samples were obtained using C18 as sorbent with recoveries between 81 and 110%, and RSD lower than 13.4%.

Method LOD values were below the maximum levels permitted for pesticides in surface waters, which are of 1 µg L^{-1} for individual pesticides and of 5 µg L^{-1} for total pesticides in water that will be destined for human consumption after treatment (Kuster et al., 2006). The analytical methods were validated and demonstrated to be accurate and efficient for the quantification of the selected herbicides.

4.2 Persistence and half-life of herbicides in rice paddy water

Table 2 shows information about the analytical methods, recommended dose of the active ingredient applied in the rice fields, the theoretical initial concentration in irrigation water and the persistence and half-life ($t_{1/2}$) of the herbicides in rice paddy water.

The sampled paddy water presented in general pH values ranging from 5.5 to 6.5 and the air temperature during the experiments ranged from 9 to 38 °C.

Herbicides	LODm (µg L⁻¹)	Recovery (%)	RSD (%)	Recommended dose (g a.i. ha⁻¹) [a, c]	Detected until (days)	$t_{1/2}$ (days)	References
Bentazon	0.1	89 - 107	<7.5	960 POST	20	2.1	Primel, 2003
				960 POST	22	3.8	Machado, 2003
Bispyribac sodium	0.1	93 - 99	<7.5	50 POST	35	2.4	Kurz et al., 2007
				50 POST	21	2.0	Kurz et al., 2009
				50 POST	60	12.4	Reimche, 2010
Clomazone	0.05	91 - 105	<13.4	500 POST	24	3.5	Primel, 2003
				500 POST	28	3.9	Machado, 2003
				500 POST	35	2.1	Gonçalves, 2007
				500 POST	42[b]	5[b]	Santos, 2008
				500 POST	31		Reimche, 2008
				500 POST	25	1.9	Reimche, 2010
Imazapic	0.1	81 - 106	<9.0	25 PRE+25 POST	24	16.1	Gonçalves, 2007
				25 PRE	17	12.3	Gonçalves, 2007
				25 POST	14	3,9	Gonçalves, 2007
				25 POST	39	10.2	Reimche, 2010
Imazethapyr	0.1	89 - 107	<7.1	75 PRE+75 POST	28	7.1	Gonçalves, 2007
				75 PRE	19	5.0	Gonçalves, 2007
				75 POST	14	3.1	Gonçalves, 2007
				75 PRE	13	1,6	Santos, 2008
				75 POST	20	5.2	Santos, 2008
				75 POST	28	4.5	Reimche, 2010
Metsulfuron-methyl	0.1	87 - 106	<11.7	2 POST	7	1.4	Machado, 2003
Penoxsulam	0.2	82 - 115	<8.1	48 POST	60	12.4	Reimche, 2010
Propanil	0.05	92 - 98	<13.1	3600 POST	5	0.8	Primel, 2003
				3600 POST	8	0.7	Machado, 2003
				3600 POST	10	0.5	Reimche, 2008
Quinclorac	0.2	89 - 110	<5.1	375 POST	20	2.2	Primel, 2003
				375 POST	21	3.7	Machado, 2003
				375 POST	42	12.2	Peixoto, 2007
				375 POST	84	15.5	Reimche, 2010
2,4-D	0.1	86 - 101	<12.1	200 POST	12	1.4	Primel, 2003
				200 POST	12	3.3	Machado, 2003

Table 2. Figures of merit of the analytical methods, herbicide concentrations, persistence and half-life ($t_{1/2}$) in rice paddy water. [a]The values corresponding also the theoretical initial concentration, in µg L⁻¹, in irrigation water, calculated in 10 cm high water slide; [b]Applied dose: 1500 g a.i. ha⁻¹; [c]Pre- or Post-emergence application.

Considering the different field experiments conducted in several years at the same place we can point out that the results for each herbicide are similar, but different compounds present great difference related to the persistence and half-life in paddy water. The average persistence, in days, in decreasing order was quinclorac (48.7), bispyribac sodium (38.7), clomazone (30.8), imazapic (23.5), imazethapyr (20.3), bentazon (20.0), penoxsulam (12.4), 2,4-D (12.0) and propanil (7.5). The average half-life in paddy water in decreasing order was imazapic (10.6), quinclorac (10.0), bispyribac sodium (5.6), imazethapyr (4.4), clomazone (3.3), bentazon (2.1), 2,4-D (1.4), metsulfuron-methyl (1.4) and propanil (0.8).

From the results presented it can be stated that rice paddy water need to be retained for different periods in the field after herbicide application depending of the herbicides used. The highest herbicides concentrations in water occur on the first days after the herbicide application decreasing with time, varying between the herbicides and years. The observance of the herbicides characteristics permit to reduce the water bodies pollution. Some herbicides were still detected after 30 days of the application. To minimize the risks of pollution of rivers and other water sources, it is important that the water containing pesticide residues remains in paddy for enough time to total herbicide dissipation.

In general, the herbicides concentrations found showed consistency with the ones reported in literature (Capri et al., 1999; Cumming et al., 2002; Quayle, 2003; Ross et al., 1989). So, the shorter the period between the herbicides application and the paddy water runoff, the higher the herbicides concentration in water and the potential risk to non-target organisms.

It is important to highlight that pesticide dissipation in irrigation water must be analyzed carefully because factors like soil preparation intensity, the period of time between the soil preparation and the rice broadcast, herbicide application time, water management, soil texture and soil chemical properties have a remarkable influence on herbicides persistence. The herbicides maintenance in rice paddy, through an irrigation management that reduces the water escape is important to avoid or to minimize herbicides pollution in water bodies.

For bentazon the results show that on average of three years, there was a decreasing of up to 64% of the herbicide concentration on the first seven days of the herbicide application. The average half-life of this herbicide was 2.1 days, with persistence of 20 days, being similar to the values reported by Crosby (1987), which mentioned that the concentration in paddy water fell to 22 μg L^{-1} in 6 days and was undetectable within 12 days.

In the conducted dissipation study, residues of bispyribac-sodium were found up to 60 days after application and the concentration at this time was of 0.3 μg L^{-1}. The average half-life of this herbicide was 5.6 days, with values between 2.0 and 12.4. These results corroborate Sanchez & Tarazona (2006) who reported that the dissipation of bispyribac-sodium in soil is rapid, with half-life of 2 to 7.6 days, while the dissipation in water is highly variable, with half-life of 7.7 to 56 days. Bispyribac-sodium belongs to the toxicological class II, considered very toxic.

Clomazone is a quite persistent herbicide being detected in average up to 30 days, in concordance with the results obtained by Cumming et al. (2002). Clomazone was also the most frequently found herbicide in irrigation water in other studies (Quayle, 2003). The characteristic of this herbicide results in the maintenance of high concentration of clomazone in the rice field enhancing the possibility of environmental pollution. This is very important, since studies conducted with fishes had demonstrated short-term effects of exposure to environmentally relevant concentrations of clomazone on AChE activity in brain and muscle tissue (Crestani et al., 2007; Miron et al., 2005).

For the herbicide imazapic, the persistence in paddy water ranged from 14 to 39 days, with an average value of 23 days. The average half-life of this herbicide was 10.6, with values between 3.9 and 16.1 days. The higher values were observed when the herbicide was applied in pre- and also in post-emergence.

Related to the herbicide imazethapyr, it presented a middle persistence in paddy water, with detectable residues from 13 to 28 days and an average persistence of 20.3 days. The half-life ranged from 1.6 to 7.1, with average value of 4.4 days. The difference can be related to the application form of the commercial herbicide, which can be in pre- and post-emergence, or only pre- or post-emergence. Similar results were reported by Marcolin et al. (2003), which found detectable concentrations of imazethapyr in water up to 30 days after application. Application in pre- and post-emergence had the highest half-life between the different application forms. For imazethapyr, photolysis is a major mechanism for its dissipation in anaerobic.conditions, since the microbial degradation of the herbicide in these conditions is almost negligible (Senseman, 2007). Ávila (2005) also states that, when applied in pre-emergence, the herbicide has more time for sorption to the soil, reducing its availability in the soil solution. This can affect the photodecomposition of imazethapyr applied. Study presented by Santos et al. (2008) shows that imazethapyr half-life in paddy water varied between 1.6 days, for application at the recommended dose in pre-emergence, and 6.2 days, for application in post-emergence.

Silva et al. (2009) conducted a monitoring study in surface water of rice production areas in seven regions of southern Brazil associated with the rice cropping and stated that imazethapyr, carbofuran and fipronil were detected in all regions studied.

Martini et al. (2011) conducted a study about the imazethapyr + imazapic leaching in lowland soil related to different rice irrigation managements. The herbicides are persistent and mobile in soil, and thus, management practices can affect its dynamics. Soil samples were collected from a field experiment submitted to different rice irrigation managements. Samples were sliced at 5 cm intervals up to 30 cm in depth. The bioassay compared the growth of non-tolerant rice plants grown in soil subjected to the treatments. The herbicide is concentrated at 5-20 cm depth, 134 days after the product is applied in lowland soil.

Moraes et al. (2011) investigated the toxicological responses of *Cyprinus carpio* after exposure in paddy water containing imazethapyr and imazapic, applied at the recommended dose. After 30 days in rice field, brain AChE activity decreases and in muscle it was enhanced, pointing out short- and long-term effects of these herbicides on this fish.

The herbicide metsulfuron-methyl was persistent in paddy water for 7 days, with half-life of 1.4. The recommended dose (2 g i.a. ha^{-1}) is very low resulting in an initial theoretical concentration of 2 µg L^{-1}. Pretto et al. (2011) studied the effects of commercial formulation containing metsulfuron-methyl on acetylcholinesterase (AChE), antioxidant profile and metabolic parameters in teleost fish (*Leporinus obtusidens*). This study pointed out long-term effects of exposure to commercial formulations containing metsulfuron-methyl on metabolic and enzymatic parameters.

Penoxsulam presented half-life of 12.4 days with relatively high remained concentration for several weeks. Detectable herbicide concentration was observed up to 60 days after the application of the recommended dose in post-emergence. The characteristic of this compound results in the maintenance of high concentration of penoxsulam in the rice field enhancing the possibility of environmental pollution.

The penoxsulam is rapidly adsorbed by the soil, except at pH above 8.0 (Senseman, 2007). In flooded soil, the penoxsulam occurs almost exclusively in the dissociated anionic form, but is somewhat persistent in water (Senseman, 2007). The dissipation of penoxsulam is rapid (Jabusch & Tjeerdema, 2006) and occurs mainly by microbial degradation or photolysis (Senseman, 2007).

The herbicide propanil showed the shortest persistence, with detectable residues up to the tenth day after application and half-life of 0.8 days. The rapid reduction of the concentration of propanil from 3600 to 0.1 μg L^{-1} is due to its fast hydrolysis (Barceló et al., 1998). However, propanil and its metabolite 3,4-dichloroaniline (3,4-DCA) can constitute a risk for surface waters and for human health (Pereira & Hostettler, 1992; Pastorelli et al. 1998). Monitoring studies of surface water carried out in the USA showed that 3% of the 1560 analyzed samples contained propanil in a concentration of up to 2 μg L^{-1} and 3,4-DCA was detected in 50% of the samples with concentration of up to 8,9 μg L^{-1} (EPA, 2006). In our studies, the concentration of 3,4-DCA increases until the 2nd day and then starts to decrease. The 3,4-DCA is detected up until the second week after the application of propanil and on the 14th day after the application the presence of 3,4-DCA was still detected at a concentration of 1.6 μg L^{-1}. Studies on the persistence of propanil in irrigated rice conditions conducted by Deul et al. (1977) showed that its dissipation occurs within 24 hours and that the amount of dissipated propanil corresponds to the concentration of 3,4-dichloroaniline (DCA), indicating biological degradation of propanil to DCA.

Quinclorac presented average half-life of 10 days with relatively high remained concentration until the seventh day (232 μg L^{-1}, average) representing 39% of the applied concentration. Detectable herbicide concentrations were observed in average up to 48 days after the application. Crosby (1987) published that in USA, under field conditions, quinclorac dissipated to undetectable levels in 31 days. Vencill (2002) reported that quinclorac present variable mobility depending on soil type and organic matter and it can persist in the soil for one year affecting susceptible crops in rotation programs.

For the herbicide 2,4-D, the persistence in paddy water was 12 days, varying the concentration from 2 to 50 μg L^{-1} at the end of the first week. The average half-life of this herbicide was 1.4 days. In water, the speed of 2,4-D degradation is fast depending on: concentration of nutrients, sediments, dissolved organic carbon and water oxygenation (Sanches-Brunete et al., 1991). Under simulated conditions, studies showed that light is also an important element on 2,4-D degradation, showing that this herbicide under light degradated faster than quinclorac (Lavy et al., 1998).

5. Conclusions

The information obtained with the field experiments allows a better understanding of the behavior of the herbicides in rice paddy water. The sample preparation by SPE followed by the analysis by LC-DAD or LC-MS/MS has proven to be efficient to show the decrease of the herbicides concentrations after the application of the commercial formulations at the field. LC-MS/MS permits a better quantification with confirmation of the compounds. The developed methods presented method limit of detection between 0.05 and 0.2 μg L^{-1} permitting adequate analyses of the paddy water samples from the diverse field experiments conducted in a period of several years. No interferences from the matrix was observed during the quantification of the herbicide residues.

Analyzing the results obtained, it can be recommended that irrigation water should be maintained, after application of the selected herbicides, for at least 20 days for the most herbicides and more than 30 days when more persistent herbicides, like quinclorac, bispiribac-sodium, clomazone and penoxsulam are used. This time, before releasing the water into the environment, is very important to reduce the pollution of water courses with herbicide residues. From the experiments repetitions it can be also concluded that the climatic conditions are very important in the process of decrease of the concentration in rice paddy water. Also the storage of rainwater, through irrigation managements, is very important to reduce the runoff of water from the paddy fields and the mass of pesticide transported to the environment.

6. Acknowledgments

Authors are grateful to the financial support and fellowships from CNPq (Brazil) through the projects MCT/CNPq/CT-HIDRO 01-2003 Process 503604/2003-8, MCT/CNPq/CT-HIDRO/SEAP 035-2007 Process 552546/2007-0 and MCT/CNPq 15-2007 Process 482578/2007-6. The authors thank all the collaborators involved with the conduction of the studies.

7. References

Alister, C. & Kogan, M. (2005). Efficacy of imidazolinone herbicides applied to imidazolinone-resistant maize and their carryover effect on rotational crops. *Crop Protection*, Vol. 24, No. 4, 375-379. ISSN 02612194.

Añasco, N.C.; Koyama, J. & Uno, S. (2010). Pesticide residues in coastal waters affected by rice paddy effluents temporarily stored in a wastewater reservoir in southern Japan. *Archives of Environmental Contamination and Toxicology*, Vol. 58, No. 2, 352-360, ISSN 1432-0800.

Añasco, N.C.; Uno, S.; Koyama, J.; Matsuoka, T. & Kuwahara, N. (2010). Assessment of pesticide residues in freshwater areas affected by rice paddy effluents in Southern Japan. *Environmental Monitoring and Assessment*, Vol. 160, No. 1-4, 371-83, ISSN 1573-2959.

Avila, L.A. de (2005). Imazethapyr: red rice control and resistance, and environmental fate. Doctoral dissertation, Texas A&M University, Stanford-USA. 81 p.

Baird, C. & Cann, M. (2005). Environmental Chemistry, 3rd ed., W. H. Freeman, ISBN 0716748770, New York.

Barceló, D. & Hennion, M.C. (1997). *Trace Determination of Pesticides and their Degradation Products*. Elsevier Science B.V, ISBN 9780444818423, Amsterdam.

Baughman, T.A. & Shaw, D.R. (1996). Effect of wetting/drying cycles on dissipation of bioavailable imazaquin. *Weed Science*, Vol. 44, No. 2, 380-382. ISSN 1550-2759.

Baugros, J.B.; Giroud, B.; Dessalces, G.; Grenier-Loustalot, M. & Cren-Olivé, C. (2008). Multiresidue analytical methods for the ultra-trace quantification of 33 priority substances present in the list of REACH in real water samples, *Analityca Chimica Acta*, Vol. 607, No. 28, 191-203, ISSN 0003-2670.

Bortoluzzi, E.; Rheinheimer, D.; Gonçalves, C.; Pellegrini, J.; Zanella, R. & Copetti, A. (2006). Contaminação de águas superficiais por agrotóxicos em função do uso do solo numa microbacia hidrográfica de Agudo, RS, *Revista Brasileira de Engenharia Agrícola e Ambiental*, Vol. 10, No. 4, 881-887, ISSN 1807-1929.

Bortoluzzi, E.; Rheinheimer, D.; Gonçalves, C.; Pellegrini, J.; Maroneze, A.; Kurz, M.; Bacar, N. & Zanella, R. (2007). Investigation of the occurrence of pesticide residues in rural wells and surface water following application to tobacco, *Química Nova*, Vol. 30, No. 8, 1872-1876, ISSN 0100-4042.

Bouman, B.A.M.; Castañeda, A.R.; Bhuiyan, S.I. (2002). Nitrate and pesticide contamination of groundwater under rice-based cropping systems: past and current evidence from the Philippines. *Agriculture Ecosystem & Environment*, Vol. 92, No. 2-3, 185-199, ISSN 0167-8809.

Boivin, A.; Amellal, S.; Schiavon, M. & van Genuchten, M.T. (2005). 2,4-dichlorophenoxyacetic acid (2,4-D) sorption an degradation dynamics in three agricultural soils. *Environmental Pollution*, Vol. 138, No. 1, 92-99, ISSN 0269-7491.

Cabrera, L.; Costa, F. & Primel, E. (2008). Estimativa de risco de contaminação das águas por pesticidas na região sul do estado do RS, *Química Nova*, Vol. 31, No. 8, 1982-1986, ISSN 0100-4042.

Caldas, S.; Demoliner, A. & Primel, E. (2009). Validation of a Method using Solid Phase Extraction and Liquid Chromatography for the Determination of Pesticide Residues in Groundwaters, *Journal of the Brazilian Chemical Society*, Vol. 20, No. 1, 125-132, ISSN 0103-5053.

Caldas, S.; Demoliner, A.; Costa, F.; D'Oca, M. & Primel, E. (2010) Pesticide Residue Determination in Groundwater using Solid-Phase Extraction and High-Performance Liquid Chromatography with Diode Array Detector and Liquid Chromatography-Tandem Mass Spectrometry, *Journal of the Brazilian Chemical Society*, Vol. 21, No. 4, 642-650, ISSN 0103-5053.

Caldas, S.S.; Zanella, R. & Primel, E.G. (2011). Risk Estimate of Water Contamination and Occurrence of Pesticides in the South of Brazil, In: *Herbicides and Environment*, Andreas Kortekamp, (Ed.), 471-492, InTech, ISBN 978-953-307-476-4, Rijeka, Croatia.

Capri, E.; Cavanna, S.; Trevisan, M. (1999). Ground and surface water bodies contamination by pesticides use in paddy field. In: *Environmental risk parameters for use of plant protection products in rice*, Capri E. et al. (Ed.), 48-71, Tipolitografia, Piacenza.

Carabias-Martínez, R.; Rodríguez-Gonzalo, E.; Herrero-Hernández, E.; Román, F. & Flores, M. (2002). Determination of herbicides and metabolites by solid-phase extraction and liquid chromatography Evaluation of pollution due to herbicides in surface and groundwaters, *Journal of Chromatography A*, Vol. 950, No. 1-2, 157–166, ISSN 0021-9673.

Carabias-Martínez, R.; Rodríguez-Gonzalo, E.; Fernández-Laespada, M.E.; Calvo-Seronero, L. & Sánchez-San Román, F.J. (2003). Evolution over time of the agricultural pollution of waters in an area of Salamanca and Zamora (Spain). *Water Research*, Vol. 37, No. 4, 928-938, ISSN 0043-1354.

Cerejeira, M.; Viana, P.; Batista, S.; Pereira, T.; Silva, E.; Valério, M.; Silva, A.; Ferreira, M. & Silva-Fernandes, A. (2003). Pesticides in Portuguese surface and ground Waters, *Water Research*, Vol. 37, No. 5, 1055–1063, ISSN 0043-1354.

Comoretto, L.; Arfib, B.; Talva, R.; Chauvelon, P.; Pichaud, M.; Chiron, S. & Höhener, P. (2008). Runoff of pesticides from rice fields in the Ile de Camargue (Rhône river delta, France): Field study and modeling. *Environmental Pollution*, Vol. 151, No. 3, 486-493, ISSN 02697491.

CONAB (2010). Séries históricas: grãos. August 2010. Available from: http://www.conab.gov.br/conteudos.php?a=1252&t=2.

Correa, I.E. & Steen, W.C. (1995). Degradation of propanil by bacterial isolates and mixed populations from a pristine lake. *Chemosphere*, Vol. 30, No. 1, 103-116, ISSN 0045-6535.

Coupe, R.H.; Thurman, E.M. & Zimmerman, R. (1998). Relation of usage to occurrence of cotton and rice herbicides in three streams of the Mississippi Delta. *Environmental Science and Technology*, Vol. 32, No. 23, 3673-3680. ISSN 1520-5851.

Crestani, M.; Menezes, C.; Glusczak, L.; Miron D.S.; Spanevello, R.; Silveira, A.; Gonçalves, F.F.; Zanella, R.; Loro, V.L. (2007). Effect of clomazone herbicide on biochemical and histological aspects of silver catfish (Rhamdia quelen) and recovery pattern. *Chemosphere*, Vol. 67, No. 11, 2305-2311, ISSN 0045-6535.

Crosby, D.R. (1987). Environmental fate of pesticides-87. California Rice Research Board. Available from: http://www.syix.com/rrb/87rpt/Enviro.htm.

Cumming, J.P.; Doyle, R.B. & Brown, P.H. (2002). Clomazone dissipation in four Tasmanian topsoils. *Weed Science*, Vol. 50, No. 3, 405-409. ISSN 1550-2759.

Curran, W.S.; Liebl, R.A. & Simmons, F.W. (1992). Effects of tillage and application methods on clomazone, imazaquin, and imazethapyr persistence. *Weed Science*, Vol. 40, No. 3, 482-489, ISSN 1550-2759.

Dahchour, A.; Bitton, G.; Coste, C.M. & Bastide, J. (1986). Degradation of the herbicide propanil in distilled water. *Bulletin of Environmental Contamination and Toxicology*, Vol. 36, No. 4, 556-562, ISSN 1432-0800.

Demoliner, A.; Caldas, S.; Costa, F.; Gonçalves, F.F.; Clementin, R.; Milani, M. & Primel, E.G. (2010). Development and Validation of a Method Using SPE and LC-ESI-MS-MS for the Determination of Multiple Classes of Pesticides and Metabolites in Water. *Journal of the Brazilian Chemical Society*, Vol. 21, No. 8, 1-10, ISSN 0103-5053.

Deul, L.E.; Brown, K.W.; Turner, F.C.; Westfall, D.G. & Price, J.D. (1977). Persistence of propanil, DCA, and TCAB in soil and water under flooded rice culture. *Journal of Environmental Quality*, Vol. 6, No. 2, 127-132, ISSN 1537-2537.

Dores, E.F.G.C. & De-Lamonica-Freire, E.M. (2001). Aquatic environment contamination by pesticides. Case study: water used for human consumption in Primavera do Leste, Mato Grosso. *Química Nova*, Vol. 24, No. 1, 27-36. ISSN 0100-4042.

Environmental Protection Agency - EPA (1985). Chemical fact sheet for bentazon and sodium bentazon. Fact sheet no. 64. Office of Pesticide Programs. Washington, DC.

Environmental Protection Agency - EPA (2001). Bispyribac-Sodium: Pesticide Tolerance Related Material, Vol. 66, No. 181. Available from: http://www.gpo.gov/fdsys/pkg/FR-2001-09-18/html/01-23227.htm.

Environmental Protection Agency - EPA (2004). Pesticide Fact Sheet No.185; U.S. EPA Office of Prevention, Pesticides and Toxic Substances: Washington, DC.

Environmental Protection Agency - EPA (2006). Overview of propanil risk assessment. Available from: http://www.epa.gov/pesticides/reregistration/status.htm.

Gburek, W.J. & Sharpley, A.N. (1997). Hydrologic controls on phosphorus loss from upland agricultural watersheds. *Journal of Environmental Quality*, Vol. 27, No. 2, 267-277, ISSN 1537-2537.

Goetz, A.; Lavy, T. & Gbur, E. (1990). Degradation and Field persistence of imazethapyr. *Weed Science*, Vol. 38, No. 2, 421-428, ISSN 1550-2759.

Gonçalves, F.F. (2007). Study of methods using HPLC-DAD and LC-MS/MS for the determination of herbicide residues in water and soil of the irrigated rice cultivation. Doctoral thesis, Universidade Federal de Santa Maria, Brazil.

Grützmacher, D.; Grützmacher, A.; Agostinetto, D.; Loeck, A.; Roman, R.; Peixoto, S. & Zanella, R. (2008). Monitoramento de agrotóxicos em dois mananciais hídricos no sul do Brasil, *Revista Brasileira de Engenharia Agrícola e Ambiental,* Vol. 12, No. 6, 632–637, 1807-1929. ISSN 1807-1929.

Hatrík, S. & Tekel, J. (1996). Extraction methodology and chromatography for the determination of residual pesticides in water. *Journal of Chromatography A,* Vol. 733, No. 1-2, 217-233, ISSN 0021-9673.

Huber, R. & Otto, S. (1994). Environmental behavior of bentazon herbicide. *Reviews of Environmental Contamination and Toxicology,* Vol. 137, No. 1, 111-134, ISSN 0179-5953.

Huber, A.; Bach, M. & Frede, H.G. (2000). Pollution of surface waters with pesticides in Germany: modeling non-point source inputs. *Agriculture Ecosystem and Environment,* Vol. 80, 191-204, ISSN 0167-8809.

Jabusch, T.W. & Tjeerdema, R.S. (2006). Photodegradation of penoxsulam. *Journal of Agricultural and Food Chemistry,* Vol. 54, No. 16, 5958-5961, ISSN 1520-5118.

Jury, W. A., Russo, D., Streile, G. & Abd, H.E. (1990). Evaluation of volatilization by organic chemicals residing below the soil surface. *Water Resources Research,* Vol. 26, No. 1, 13-20, ISSN 0043-1397.

Kammerbauer, J. & Moncada, J. (1998). Pesticide residue assessment in three selected agricultural production systems in the Choluteca River Basin of Honduras. *Environmental Pollution,* Vol. 103, No. 2, 171-181, ISSN 0269-7491.

Kirksey, K.B.; Hayes, R. M.; Krueger, W.A. & Mueller, T.C. (1996). Clomazone dissipation in two Tennessee soils. *Weed Science,* Vol. 44, No. 4, 959-963, ISSN 1550-2759.

Kolpin, D.W.; Thurman, E.M. & Linhart, S.M. (1998). The environmental occurrence of herbicides: The Importance of degradates in ground water. *Archives of Environmental Contamination and Toxicology,* Vol. 35, No. 3, 385-390. ISSN 1432-0703.

Kolpin, D.W.; Thurman, E.M. & Linhart, S.M. (2000). Finding minimal herbicide concentration in ground water? Try looking for their degradates. *Science of the Total Environment,* Vol. 248, No. 2-3, 115-122, ISSN 0048-9697.

Konstantinou, I.; Hela, D. & Albanis, T. (2006). The status of pesticide pollution in surface waters (rivers and lakes) of Greece. Part I. Review on occurrence and levels, *Environmental Pollution,* Vol. 141, No. 3, 555-570, ISSN 0269-7491.

Kraemer, A.F.; Marchesan, E.; Avila, L.A.; Machado, S.L.O.; Grohs, M.; Massoni, P.F.S. & Sartori, G.M.S. (2009). Persistence of the herbicides imazethapyr and imazapic in irrigated rice soil. *Planta Daninha,* Vol. 27, No. 3, 581-588, ISSN 0100-8358.

Kurz, M.H.S. (2007). Study of methods using solid-ohase extraction and analysis by HPLC-DAD and GC-ECD for the determination of pesticide residues in waters and field degradation. Doctoral thesis, Universidade Federal de Santa Maria, Brazil.

Kurz, M.H.S.; Gonçalves, F.F.; Martel, S.; Adaime, M.B.; Zanella, R.; Machado, S.L. de O. & Primel, E.G.. (2009). Rapid and accurate hplc-dad method for the determination of the herbicide bispyribac-sodium in surface water, and its validation. *Química Nova,* Vol. 32, No. 6, 1457-1460, ISSN 0100-4042.

Kuster, M.; Alda, M. L. & Barceló, D. (2006) Analysis of pesticides in water by liquid chromatography-tandem mass spectrometric techniques. *Mass Spectrometry Reviews,* Vol. 25, No. 6, 900-916, ISSN 1098-2787.

Laganà, A.; Bacaloni, A.; Leva, I.; Faberi, A.; Fago, G. & Marino, A. (2002). Occurrence and determination of herbicides and major transformation products in environmental waters, *Analytica Chimica Acta,* Vol. 462, No. 2, 187-198, ISSN 0003-2670.

Lavy, T.L.; Mattice, J.D. & Norman, R.J. (1998). Environmental implications of pesticides in rice production. In: *Rice research studies*. Norman, R.J. & Johnston, T.H. (Ed.), 63-71, Arkansas Agricultural Experimental Station. Fayetteville-Arkansas.

Lee, D.J.; Senseman, S.A.; O'barr, J.H.; Chandler, J.M.; Krutz, L.J. & Mccauley, G.N. (2004). Soil characteristics and water potential effects on plant-available clomazone in rice. *Weed Science*, Vol. 52, 310-318, ISSN 1550-2759.

Loux, M.M. & Reese, K.D. (1992). Effect of soil pH on adsorption and persistence of imazaquin. *Weed Science*, Vol. 40, No. 3, 490-496, ISSN 1550-2759.

Machado, S. L. de O. (2003). Sistemas de implantação de lavouras de arroz irrigado, consumo de água, perdas de nutrientes na água de drenagem, persistência de herbicidas na água e efeitos em jundiá. Doctoral thesis, Universidade Federal de Santa Maria, Brazil.

Machado, S.L. de O.; Marchezan, E.; Righes, A.A.; Carlesso, R.; Villa, S.C.C. & Camargo, E.R. (2006). Water use and nutrients and sediments losses on the initial water drainage on flooded rice. *Ciência Rural*, Vol. 36, No. 1, ISSN 0103-8478.

Mallipudi, N.M.; Stout, S.J.; DaCunha, A.R. & Lee, A.H. (1991). Photolysis of imazapyr (AC 243997) herbicide in aqueous media. *Journal of Agricultural and Food Chemistry*, Vol. 39, No. 2, 412-417, ISSN 1520-5118.

Marchesan, E.; Meneghetti, G.; Sartori, S.; Avila, L.; Machado, S.; Zanella, R.; Primel, E.; Macedo, V. & Marchezan, M. (2010). Resíduos de agrotóxicos na água de rios da Depressão Central do Estado do Rio Grande do Sul, Brasil, *Ciência Rural*, Vol. 40, No. 5, 1053-1059, ISSN 0103-8478.

Marchesan, E.; Zanella, R.; Avila, L.; Camargo, E.; Machado, S. & Macedo, V. (2007). Rice herbicide monitoring in two brazilian rivers during the rice growing season, *Scientia Agricola*, Vol. 64, No. 2, 131-137, ISSN 0103-9016.

Marcolin, E.; Macedo, V.R.M. & Genro Junior, S.A. (2003). Persistência do herbicida imazethapyr na lâmina de água em três sistemas de cultivo de arroz irrigado, *Proceedings of the Brazilian Rice Meeting*, pp. 686-688. Camboriú, Epagri, Camboriú, SC, Brazil.

Martini L.F.D.; Avila, L.A.; Souto, K.M.; Cassol, G.V., Refatti, J.P.; Marchesan, E. & Barros, C.A.P. (2011). Imazethapyr + imazapic leaching in lowland soil as affected by rice irrigation management. *Planta Daninha*, Vol. 29, No.1, 185-193, ISSN 0100-8358.

McPherson, A. K.; Moreland, R. S.; Atkins, J. B. (2003). Occurrence and Distribution of Nutrients, Suspended Sediment, and Pesticides in the Mobile River Basin, Alabama, Georgia, Mississippi, and Tennessee, 1999-2001. Water-Resources Investigations Report 03-4203, U.S. Geological Survey: Montgomery, pp 1-2, 44, 57.

Mervosh, T.L.; Simms, G.K.; Stoller, E.W. (1995). Clomazone fate as affected by microbial activity, temperature, and soil moisture. *Journal of Agricultural and Food Chemistry*, Vol. 43, No. 2, 537-543, ISSN 1520-5118.

Miron, D.S.; Crestani, M.; Shettinger, R. M; Morsch, V. M.; Baldisserotto, B.; Tierno, M. A.; Moraes, G.; Vieira, V. L. (2005). Effects of the herbicides clomazone, quinclorac, and metsulfuron methyl on acetylcholinesterase activity in the silver catfish (*Rhamdia quelen*) (Heptapteridae). *Ecotoxicology and Environmental Safety*, Vol. 61, No. 3, 398-403, ISSN: 0147-6513.

Moraes, B.S.; Clasen, B.; Loro, V.L.; Pretto, A.; Toni, C.; Avila, L.A.; Marchesan, E.; Machado, S.L.O.; Zanella, R. & Reimche, G.B. (2011). Toxicological responses of *Cyprinus carpio* after exposure to a herbicide containing imazethapyr and imazapic. *Ecotoxicology and Environmental Safety*. Vol. 74, No. 3, 328-335, ISSN: 0147-6513.

Nakano, Y.; Miyazaki, A.; Yoshida, T.; Ono, K. & Inoue, T. (2004). A study on pesticide runoff from paddy fields to a river in rural region - 1: field survey of pesticide runoff in the Kozakura River, Japan. *Water Research*, Vol. 38, No. 13, 3017-3022, ISSN 0043-1354.

Noldin, J.A.; Ederhardt, D.S.; Deschamps, F.C. & Hermes, L.C. (2001). Strategies for water sampling for monitoring the rice enviromental impact. *Proceedings of the Brazilian Rice Meeting*, pp.760-762, Porto Alegre, 2001, IRGA, Porto Alegre, Brazil.

Noldin, J.A.; Hermes, L.C.; Rossi, M.A. (1997). Persistence of clomazone in pre-germinated rice paddy water. *Proceedings of the Brazilian Rice Meeting*, pp.363-364, Camboriú, SC, Brazil.

Numabe A. & Nagahora, S. (2006). Estimation of pesticide runoff from paddy fields to rural rivers. *Water Science and Technology*, Vol. 53, No. 2, 139-146, ISSN 0273-1223.

Palma, G.; Sánchez, A.; Olave, Y.; Encina, F.; Palma, R. & Barra, R. (2004). Pesticide levels in surface waters in an agricultural–forestry basin in Southern Chile, *Chemosphere*, Vol. 57, No. 8, 763-770. ISSN 0045-6535.

Pastorelli, R.; Catenacci, G.; Guanci, M.; Fanelli, R.; Valoti, E.; Minoia, C. & Airoldi, L. (1998). 3,4-Dichloroaniline-haemoglobin adducts: Preliminary data on agricultural workers exposed to propanil. *Biomarkers*, Vol. 3, No. 3, 227-233, ISSN 1366-5804.

Peixoto, S.C. (2007). Study of field pesticide stability of carbofuran and quinclorac in the water of irrigated rice crops using SPE and HPLC-DAD. Master dissertation, Universidade Federal de Santa Maria, Brazil.

Pereira, W.E. & Hostettler, F.D. (1992). Nonpoint source contamination of the Mississippi River and its tributaries by herbicides. *Environmental Science and Technology*, Vol. 27, No. 8, 1542-1552, ISSN 1520-5851.

Pereira, W.E. & Rostad, C.E. (1990). Occurrence, distributions, and transport of herbicides and their degradation products in the lower Mississippi River and its tributaries. *Environmental Science and Technology*, Vol. 24, No. 9, 1400-1406, ISSN 1520-5851.

Pretto, A.; Loro, V.L.; Menezes, C.; Moraes, B.S.; Reimche, G.B.; Zanella, R.; de Avila, L.A. (2011). Commercial formulation containing quinclorac and metsulfuron-methyl herbicides inhibit acetylcholinesterase and induce biochemical alterations in tissues of *Leporinus obtusidens*. *Ecotoxicology and Environmental Safety*, Vol. 74, No. 3, 336-341, ISSN: 0147-6513.

Primel, E.G. (2003). Application of solid-phase extraction and chromatographic techniques for the determination of herbicides in surface water and accompaniment of the degradation in field and in laboratory. Doctoral thesis, Universidade Federal de Santa Maria, Brazil.

Primel, E.G.; Zanella, R.; Kurz, M.H.S.; Gonçalves, F.F.; Machado, S.L.O. & Marchezan, E. (2005). Pollution of water by herbicides used in the irrigated rice cultivation in the central area of Rio Grande do Sul state, Brazil: theoretical prediction and monitoring. *Química Nova*, Vol. 28, No. 4, 605-609, ISSN 0100-4042.

Quayle, W.C. (2003). Persistence of rice pesticides in floodwaters: influence of water management. *Proceedings of the 3rd International Temperate Rice Conference*, pp. 97, Vol. 3, Punta del Este, Uruguay, INIA.

Reimche, G.B. (2010). Impacto de agroquímicos usados na lavoura de arroz irrigado sobre a qualidade da água de irrigação e na sobre a comunidade zooplanctônica. Master dissertation, Universidade Federal de Santa Maria, Brazil.

Reimche, G.B.; Machado, S.L.O.; Golombieski, J.I.; Baumart, J.S.; Braun, N.; Marchesan, E. & Zanella, R. (2008). Water persistence and influence of herbicides utilized in rice paddy about zooplankton community of Cladocers Copepods and Rotifers. *Ciência Rural*, Vol. 38, No.1, 7-13, ISSN 0103-8478.

Resgalla, C.Jr; Noldin, J.A.; Tamanaha, M.S.; Deschamps, F.C.; Eberhardt, D.S. & Rörig L.R. (2007). Risk analysis of herbicide quinclorac residues in irrigated rice areas, Santa Catarina, Brazil. *Ecotoxicology*, Vol. 16, No. 8, 565-571, ISSN 1573-3017.

Ross, L.J.; Powell, S. and Fleck, S.L. (1989). Dissipation of bentazon in flooded rice field. *Journal of Environmental Quality*, Vol. 18, No. 1, 105-109, ISSN 1537-2537.

Sanches-Brunete, C.; Perez, S. & Tadeo, J.L. (1991). Determination of phenoxy ester herbicides by gas and high-performance liquid chromatography. *Journal of Chromatography A*, Vol. 552, No. 1-2, 235-240, ISSN 0021-9673.

Santos, F.M.; Marchesan, E.; Machado, S.L.O.; Avila, L.A.; Zanella, R. & Gonçalves, F.F. (2008). Imazethapyr and clomazone persistence in rice paddy water. *Planta Daninha*, Vol. 26, No. 4, 875-881, ISSN 0100-8358

Santos, T.C.R., Rocha, J.C., Barceló, D. (2000). Determination of rice herbicides, their transformation products and clofibric acid using on-line solid-phase extraction followed by liquid chromatography with diode array and atmospheric pressure chemical ionization mass spectrometric detection. *Journal of Chromatography A*, Vol. 879, No. 1, 3-12, ISSN 0021-9673.

Santos, T.C.R.; Rocha, J.C.; Alonso, R.M.; Martinez, E.; Ibanez, C. & Barceló, D. (1998). Rapid degradation of propanil in rice crop fields. *Environmental Science and Technology*, Vol. 32, No. 22, 3479-3484, ISSN 1520-5851.

Senseman, S.A. (Ed.) (2007). *Herbicide handbook*. 9th ed., Lawrence: Weed Science Society of America. ISBN: 978-1891276569, Lawrence, KS. 458 pp.

Siemering, G.S., Hayworth, J.D. & Greenfield, B.K. (2008). Assessment of potential aquatic herbicide impacts to California aquatic ecosystems. *Archives of Environmental Contamination and Toxicology*, Vol. 55, No. 3, 415-31, ISSN 1432-0703.

Silva, D.; Avila, L.; Agostinetto, D.; Dal Magro, T.; Oliveira, E.; Zanella, R. & Noldin, J. (2009). Monitoramento de agrotóxicos em águas superficiais de regiões orizícolas no sul do Brasil, *Ciência Rural*, Vol. 39, No. 9, 2383-2389, ISSN 0103-8478.

Solomon, K.R.; Baker, P.R.; Dixon, K.R.; Klaine, S.J.; LaPoint, T.W.; Kendall, R.J.; Weisskopf, C.P.; Giddings, J.M.; Giesy, J.P.; Hall, L.W. & Williams, W.M. (1996). Ecological risk assessment of atrazine in North American surface waters. *Environmental Toxicology and Chemistry*, Vol. 15, No. 1, 31-76, ISSN 1552-8618.

Son, H.V.; Ishihara, S. & Watanabe, H. (2006). Exposure risk assessment and evaluation of the best management practice for controlling pesticide runoff from paddy fields. Part 1: Paddy watershed monitoring. *Pest Management Science*, Vol. 62, No. 12, 1193-1206, ISSN: 1526-4998.

Squillace, P.J. & Thurmann, E.M. (1992). Herbicide transport in rivers: Importance of hydrology and geochemistry in nonpoint-source contamination. *Environmental Toxicology and Chemistry*, Vol. 26, No. 3, 538-545, ISSN 1552-8618.

Steckert, A.V.; Schnack, C.E.; Silvano, J.; Dal-Pizzol, F. & Andrade, V.M. (2009). Markers of pesticide exposure in irrigated rice cultures. *Journal of Agricultural and Food Chemistry*, Vol. 57, No. 23, 11441-11445, ISSN 1520-5118.

Stougaard, R.N.; Shea, P.J. & Martin, A.R. (1990). Effect of soil type and pH on adsorption, mobility and efficacy of imazaquin and imazethapyr. *Weed Science*, Vol. 36, No. 1, 67-73, ISSN 1550-2759.

Sudo, M.; Kunimatsu, T. & Okubo, T. (2002). Concentration and loading pesticides residues in Lake Biwa Basin (Japan). *Water Research*, Vol. 36, No. 1, 315-329, ISSN 0043-1354.

Tarazona, J.V. & Sanchez, P. (2006). Development of an innovative conceptual model and a tiered testing strategy for the ecological risk assessment of rice pesticides. *Paddy and Water Environment*, Vol. 4, No. 1, 53-59, ISSN 1611-2504.

Thurman, E.M.; Goolsby, D.A.; Meyer, M. T. & Kolpin, D.W. (1991). Herbicides in surface waters of the Midwestern United States: The effect of spring flush. *Environmental Science and Technology*, Vol. 25, No. 10, 1794-1796, ISSN 1520-5851.

Thurman, E.M.; Goolsby, D.A.; Meyer, M. T.; Mills, M.S.; Pomes, M.I. & Kolpin, D.W. (1992). A reconnaissance study of herbicides and their metabolites in surface water of the Midwestern United States using immunoassay and gas chromatography/mass spectrometry. *Environmental Science and Technology*, Vol. 26, No. 12, 2440-2447, ISSN 1520-5851.

Tomlin, C.D.S. (2007). *A world compendium: the e-Pesticide Manual*, 14th ed. CDROM version 4.0, The British Crop Protection Councill, Farnham, UK. ISBN 1-901396-42-8

Ueji, M. & Inao, K. (2001). Rice paddy field herbicides and their effects on the environment and ecosystems. *Weed Biology and Management*, Vol. 1, No. 1, 71–79, ISSN 1445-6664.

U.S. Department of Agriculture – USDA (2007). Pesticide Data Program Annual Summary, Calendar Year 2006, Agricultural Marketing Service: Washington, DC.

Vencill, W. K. (2002). *Herbicide Handbook*, 8th ed., Weed Science Society of America, ISBN 1891276336, Lawrence, KS.

Walls, D.; Smith, P.G. & Mansell, M.G. (1996). Pesticides in groundwater in Britain. *International Journal of Environmental Health Research*, Vol. 6, No. 1, 55-62, ISSN 1369-1619.

Ware, G. W. (1994). *The Pesticide Book*, 4th ed., Thomson Publications, ISBN: 0913702587, Fresno, CA.

Woudneh, M.B.; Ou, Z.; Sekela, M.; Tuominen, T. & Gledhill, M. (2009). Pesticide multiresidues in waters of the Lower Fraser Valley, Canada. Part I. Surface water. *Journal of Environmental Quality*, Vol. 8, No. 3, 940-947, ISSN: 00472425.

Zanella, R.; Primel, E.G.; Gonçalves, F.F.; Kurz, M.H.S. & Mistura, C.M. (2003) Development and validation of a high-performance liquid chromatographic procedure for the determination of herbicide residues in surface and agriculture waters. *Journal of Separation Science*, Vol. 26, No. 9-10, 935-938, ISSN 1615-9314.

Zanella, R.; Primel, E.G.; Machado, S.L.O.; Gonçalves, F.F. & Marchezan, E. (2002). Monitoring of the Herbicide Clomazone in Environmental Water Samples by Solid-Phase Extraction and High-Performance Liquid Chromatography with Ultraviolet Detection, *Chromatographia*, Vol. 55, No. 9-10, 573-577, ISSN 1612-1112.

Zanella, R.; Primel, E.G.; Gonçalves, F.F. & Martins, A.F. (2000). Development and validation of a high-performance liquid chromatographic method for the determination of clomazone residues in surface water. *Journal of Chromatography A*, Vol. 904, No. 2, 257-262, ISSN 0021-9673.

10

Gene Flow Between Conventional and Transgenic Soybean Pollinated by Honeybees

Wainer César Chiari[1], Maria Claudia Colla Ruvolo-Takasusuki[2],
Emerson Dechechi Chambó[1], Carlos Arrabal Arias[3],
Clara Beatriz Hoffmann-Campo[3] and Vagner de Alencar Arnaut de Toledo[1]
[1]*Animal Science Department, Universidade Estadual de Maringá, Maringá, Paraná*
[2]*Cell Biology and Genetics Department,*
Universidade Estadual de Maringá, Maringá, Paraná
[3]*Empresa de Brasileira de Pesquisa Agropecuária - Soja, Londrina, Paraná*
Brazil

1. Introduction

Among the main agricola commodities of Brazil, the soybean (*Glycine max* L. Merrill) keeps a highlight place. In last years, the cultivated area with this Leguminosae has been growing, as well as was testified the increase of productivity by area in Brazilian crops. The weeds are the principal problems that interferes in soybean production. After several decades of searching alternatives to control plagues of considerable periods of time, genotype of genetically modified (GM) soybean were developed to be resistant to herbicide glyphosate, an herbicide of chemical group of substituted glycine.

Wild soybean *G. max* is distributed in the Far East of Russia, eastern China including Taiwan, Korean peninsula and Japan (Lu, 2004). This annual plant grows in edges of crop fields, roadsides and riverbanks (Kuroda et al., 2010). The centre of domestication is controversial, may have occurred in China or independently in several regions of East Asia (Hymowitz, 1970, Xu et al., 2002). Only at the late 1940's and early 1950's, the U.S. exceeded China and eventually the entire Orient in soybean production (Hymowitz, 1970). The tolerance to herbicide was obtained by addiction in soybean genome of a gene named CP4, which came from common bacteria of soil of whole world *Agrobacterium* sp. The herbicide acts in aromatic chain amino acids synthesis inhibiting the enzyme EPSPs, in which the metabolic route synthesizing proteins, vitamins, hormones, and other essential products to growing and development of invading plants (Gazziero et al., 2009a). Therefore, the farmer can use the glyphosate to chemical weeding without risk to soybean plants (Paula-Júnior & Venzon, 2007). The herbicides are items of highest cost in the production system. The glyphosate in post emergence in soybean crops represents the possibility of using a new tool in management of plagues; easily handle, flexible and efficient. By presenting these features, the farmers of soybean were stimulated to adopt this technology (Gazziero et al., 2009b).

The cultivation of GM plants in 2010 overtaken, in whole world, the mark 148 million hectares, and the soybean plays 60% of total GM plants (James, 2010). The transgenic soybean cultivation was regulated by the Brazilian Government – Bio security Law from March 2005, which authorizes the production and commercialization of GM products, and set security rules and mechanisms of controlling of any activities with organisms genetically modified. Its utilization by farmer is increasing year by year more than conventional varieties (Menegatti & Barros, 2007), and this Bio security Law also regulates the research about stem cells (Bruno, 2008). Besides, the government creates a commission named National Technical Commission of Bio security (CNTBio).

Brazil increased the planted area more than any other country, with an impressive rise of four million hectares in relation to 2009, and then, Brazil is now the second position of the commercial cultivation of GM plants, 25.4 million hectares, in whole word (James, 2010). The GM soybean, in 2010, occupied 50% of cultivated global area with OGMs, reaching 73.3 million ha, followed by corn – 46.8 million ha (31%), by cotton – 21 million ha (14%), and by canola – 7 million ha (5%). Since the beginning of commercialization in 1996 until 2010, the tolerance to herbicides is the dominant attribute. In USA, in 2010, the tolerance to herbicides used in soybean, corn, canola, cotton, beetroot, and alfalfa occupied 61% or 89.3 million ha of global area of 148 million ha cultivated with biotechnological planting (James, 2010).

The green revolution gets reduce the percentage of word-wide population that suffer hunger from 50% in sixties years to 20% nowadays. In plantation free of agrotoxic, the loss of production is 10% to 40%. If it not been used this technique, about two billion of six billion of inhabitants of the planet would starving. The use of transgenic crops may help to increase the productivity, avoiding, more deforesting, and more erosion of the soils (Souza, 2006).

Despite of evaluation of risk of the release of OGMs had been made cautiously by CTNBio, following procedures that avoid or minimize the adverse consequences of OGMs and their derivates to human being and environment (Mendonça-Hagler, 2001), one of the most worries of Brazilian farmers in relation to GM plants is the unexpected and cumulative effect of cross-contamination between GM plants and the conventional. The instability and risk of propagation of a gene to wild species are critical to maintenance of the environment (Chiari et al., 2008). Although the cross-pollination provides an increase in gene flow between plants there is a diversification with results remarkably favourable (McGregor, 1976; Free, 1993; Malerbo-Souza et al., 2004; Chiari et al., 2005a; Chambó et al., 2010). The genetic pollution is inevitable, whereas the transgenic pollen may contaminate conventional or biological fields located several kilometres from GM plantation (Scottish Crop Research Institute, 1999). Therefore, questions about bio security had been discussed, including gene flow via pollen (Ray et al., 2003; Schuster et al., 2007; Silva & Maciel, 2010). Besides, the impact of cross-pollination between transgenic cultivar and conventional in assays of improvement is not known.

In an insect-pollinated crop, gene flow occurs when pollinator moves between fields and cross-pollinate flowers with 'foreign' pollen (Cresswell, 2010). The gene flow via pollen by biological agents is a primordial factor in cross-pollination rates. Even though the soybean is considered auto pollinated, or not is beneficiate of insect pollination (Rubis, 1970, Ahrent & Caviness, 1994, Wolff, 2000), some authors considered the occurrence of cross-pollination in this species attributed to action of insect pollinators (Erickson, 1975; McGregor, 1976; Chiari et al., 2005a,b). However, the cv. BRS-133, not GM (Chiari et al., 2005a,b) was intensively visited by *A. mellifera* Africanized honeybees. The researchers reported an increase of 61.38% in number of pods, and 58.86% in seeds production, when it is compared with protected plants of the insect visitation.

The possibility of gene flow occurrence of the GM variety to conventional and the possibility of detect soybean seeds resistant to glyphosate mixed with conventional soybean are relevant subject with coming and spread of OGMs. In economical globalization and expansion of transgenic in agriculture, the diversity of market requires cultivar and these products must be precisely identified to commercialization. There are several commercial niches in agribusiness, and is imperative the need of obtain tools to detect the main characteristics of cultivar, as for seeds production with genetic purity guaranteed, as for the certification of the products (Pereira, 2007).

The transgenic flow becomes a problem especially because the farmer is interested in growing organic or conventional soybeans. Contamination with the transgenic will provide damage to the farmer.

Non-transgenic cultivars could be contaminated by transgene via wind-pollination. Yoshimura (2011) investigated this potential contamination by assessment of soybean pollen dispersion. The airborne soybean pollen was sampled using Durham pollen samplers located in the range of 20 m from the field edge. The dispersal distance was assessed in a wind tunnel under constant airflow and it was compared with the anticipated distances based on the pollen diameter. It was detected little airborne pollen in and around the field and the dispersal is restricted to a small area from the field edge even when soybean flowers were in full bloom. Considering soybean characteristics with a stigma invisible from the outside and a short pollen life, wind-mediated pollination in soybeans appears to be negligible.

Coexistence among genetically modified (GM) with non-GM cropping systems and identifying preservation at the field level are increasingly important issues in many countries (Beckie & Hall, 2008). In several types of cultivated plants of economical interest, the gene flow by cross-pollination is possible between separated spatially fields (Beckie & Hall, 2008). To preserve the purity of the varieties of a crop or to restrict the introduction of genes of GM soybean in conventional crops, a better understanding about gene flow is unexpected. Besides, there is a great need to generate knowledge about OGMs so that can perform an evaluation about benefits and what their implications in the ecosystem. One aspect to be studied refers to floral biology and pollination. In reproduction of several plants, the pollination represents an important mechanism to increase the gene flow in the species.

The insects play a fundamental role in transfer these genes by cross-pollination, this process that provides equilibrium in the ecosystem. Meanwhile the insects searching floral resources like pollen and nectar, they offer to plant a diversity of genetic material. Therefore, it is necessary to consider the possibility of these OGMs provoke alterations in this equilibrium (Chiari et al., 2005b).

The importance of bee pollination in particular can be understood (Pasquet et al., 2008) in Kenya using carpenter bees *Xylocopa flavorufa* main pollinator of *Vigna unguiculata* (cowpea). The experiment was carried out with the release of an insect-resistant, genetically engineered. The authors found out bees visited wild and domesticated populations, can mediate gene flow and, in some instances, allow transgenic escape over several kilometres.

In this research was evaluated the cultivar BRS-245 RR, developed by Empresa Brasileira de Pesquisa Agropecuária (Embrapa Soybean), and it has the characteristic Roundup Ready™ (Soja RR), developed by Monsanto, compared with conventional cultivar BRS 133, a isoline of BRS 245 RR, not transgenic. The choice of this OGM was made based on the importance of the soybean crop to the state of Paraná, and for Brazil, the greatest word-wide exporter, and for the availability of the cultivar by Embrapa Soja of Londrina city. The objective of this research was evaluating the gene flow provoked by cross rate between transgenic soybean and conventional and to verify the influence of Africanized honeybees A. *mellifera* in this process on cultivars BRS 245 RR and its isoline the BRS 133.

2. Material and methods

This research was carried out in experimental area of Empresa Brasileira de Pesquisa Agropecuária (Embrapa Soja), located in Londrina city (23° 08'47" S and 51° 19'11" W), which is situated in North region of state of Paraná, Brazil. The planting season, the cultivation, and management of the culture and crop occurred in appropriated time, and followed technical recommendations to soybean plant (Embrapa, 2003).

It was used the completely randomizely design with three treatments and six repetitions each. It was evaluated three treatments: covered area with cages, and inside there was a colony of honeybees during the flowering (Figure 1); a covered area without honeybee colony; and an uncovered area, free for insect visitation. In each area, of 24 m^2 each, the soybean planting was made in eight lines, of 6 m, interlaced two by two, with cultivars BR 245 RR and BRS 133. The stand used was 0.5 m between lines and 30 seeds by linear metre (Figure 2).

In covered areas, pollination cages were installed, made with nylon screen (two mm), supported by PVC tubes (¾ inch), and iron (3/8 inch), forming cages in a semi-arch with four metres wide, six metres length and two metres high, covering an area of 24 m^2 (Figure 3) and this avoids the passage of insects (Chiari et al., 2005a).

Immediately before the blooming phase, the cages were mounted and inside of each cage of covered treatment with honeybees, it was put an Africanized colony – A. *mellifera* in a Langstroth hive model with five frames, thus three frames with brood and two with pollen and honey (Figure 1). In covered treatment without honeybees, the cage avoids the insect visitation to soybean flowers, and third treatment was a delimited area with free visitation of insects and other pollinators and or visitors (Figure 3).

The soybean plantation was monitored during all the period with particular attention to the flowering phase, which started on December 31st 2003 to January 28th 2004, where the cages were dismounted. The insect visitation was evaluated in covered area with honeybee colony inside and in uncovered area, by two individuals that tracked two lines of transgenic soybean and two lines of conventional cultivar, during 10 minutes per hour, from 8:00h a.m. to 5:00h p.m., with three repetitions.

The harvesting of soybean grains was performed separately in each line, and the borders were discarded, computing 12m^2 of each cage. Cleaning, drying, classifying, and weighing of harvested grains obtained the production. Evaluating the gene flow, 1,000 seeds harvested of each line of conventional soybean were seeded in an experimental field. After the germination, the viable seedlings were counted, and where the fourth leaf arises from

Fig. 1. A honeybee colony inside the pollination cage.

Fig. 2. Planting of transgenic and conventional soybean in alternate lines.

Fig. 3. Pollination cage model used in the assays. Measures: 4m x 2m x 6m.

seedling, it was applied glyphosate (Figure 4). The evaluation and quantification of seedlings that survived after the pulverization with glyphosate was recorded after a week.

Data were analyzed with the software Statistical Analysis System (SAS, 2004), using the following model:

$Yijklm = \mu + Bi + Hj + Tk + Vl + (BH)ij + (BT)ik + (BV)jl + (HT)jk + (HV)jl + (TV)kl + e\ ijklm.$

Wherein:

$Yijklm$ = Observation as to variable of Block i, Herbicide j, Treatment k, Cultivar l

μ is the effect of general average;
Bi is the effect of Block (i = 1, 2 ... 6);
Hj is the effect of Herbicide (j = 1, 2);
Tk is the effect of Treatment (1, 2, 3);
Vl is the effect of cultivar;
$(BH)ij$ is the interaction of Block i and Herbicide J;
$(BT)ik$ is the interaction of Block i and Treatment k;
$(BV)jl$ is the interaction of Block i and Cultivar l;
$(HT)jk$ is the interaction of Herbicide j and Treatment k;
$(HV)jl$ is the interaction of Herbicide i and Cultivar l and;
$eijklm$ the error associated to observation ijklm.

Data were analyzed by variance analysis (Anova), and means of treatments compared using Tukey's test, at level of 5% of probability.

Fig. 4. The application of glyphosate in transgenic and conventional soybean plants.

3. Results and discussion

The percentage of germination of 1,000 seeds planted in experimental field did not differ (p>0.05) between covered area with pollination cages within Africanized honeybee colony, pollination cage without honeybees inside and uncovered area for free insect visitation. In Table 1 are shown data referred to summary of analysis of variance and means of percentage of seeds germination of conventional soybean resistant to herbicide glyphosate to all treatments.

The germination rate of seeds from conventional plants, in different treatments, was 79.43% ± 18.42. This value is considered satisfactory to cultivar BRS 133 (EMBRAPA, 2003). However, the average percentage of germination obtained in this research was lower than 91.02% reported by Chiari et al. (2005a) working with the same cultivar. A possible explanation for that difference in germination rate is, while in this research this rate was measured in a field, Chiari et al. (2005b) evaluating this rate in laboratorial conditions.

The similarity in germination rate between covered area with honeybee colony, covered area without honeybee colony, and uncovered area for free insect visitation and other visitors suggests that the presence of Africanized honeybees and other insects did not interfere in germination rate of the seeds. In other species such as canola (cv. OAC Triton) was found out the presence of pollinators, e.g. honeybees, increasing the germinability of resulting seeds from 83% to 96% (Kevan & Eisikovitch, 1990).

Variation source	Germination rate (%)	
Treatment	0.51	P=0.6372
CV %	0.51	
Means of treatments	79.43 ± 18.42	
Covered area with honeybee colony	86.30	a (± 11.97)
Covered area without honeybee colony	72.83	a (± 26.15)
Uncovered area for free insect visitation	79.15	a (± 14.75)

Means followed by same small letters did not differ between them by Tukey's test (P>0.05)

Table 1. F values with their probability (P), coefficient of variation (CV%), means and their standard deviation of percentage of germination of 1,000 seeds obtained by resistant plants to glyphosate from seeds of soybean *Glycine max* L. Merrill, variety BRS 133 (conventional soybean) to treatments: covered area with honeybee colony, covered area without honeybee colony, and uncovered area for free insect visitation in experimental field of EMBRAPA Soja, in Londrina-PR city.

The environmental and physiological factors of the plant are considered the responsible directly of the better physiological quality of the seeds, including the vigour. Nevertheless, some researchers think about the insects can in several situations influence indirectly by better quality of the seeds, and then, provide seeds of the most high quality, for example, with more vigour and oil (Singh et al., 2001; Paiva et al., 2003; Camacho & Franke, 2008; Toledo et al., 2011).

The percentage of cross-pollination of plants from conventional seeds that show resistance to glyphosate in covered area with honeybee colony, covered area without honeybee colony, and uncovered area for free insect visitation are in Table 2. The percentage of transfer via pollen flow from transgenic soybean to conventional did not differ (p ≥0.05) between pollination cages with honeybee colony and uncovered area for free insect visitation (1.57%). However, it was higher (p=0.0224) than percentage of pollen transfer of the covered area without honeybee colony, and the average was 0.20%. Therefore, the Africanized honeybees and other insects were responsible by cross-pollination, increasing the gene flow from transgenic to conventional soybean. Afterward, the gene flow from conventional to transgenic soybean may also have occurred, but was not evaluated in this assay.

It was possible to estimate an increase of 685% by cross-pollination in pollen transfer from transgenic to conventional soybean plants when Africanized honeybees visited the soybean flowers in relation to these areas in which the honeybees were avoided to visit. Therefore, the Africanized honeybees are able and efficient in transferring pollen from a plant to another.

Honeybees collecting mainly nectar, while collecting nectar and pollen was observed less frequently and foraging behavior of pollen was found out only rarely (observed data, not shown). These findings show that although literature data indicate that soybean flowers are not attractive to honeybees because there is not a large quantity of nectar (Erickson, 1975; Alves et al., 2010) they visit flowers and collect nectar and/or pollen, and an increase of yield like 37.84% when honeybee visits were allowed (Chiari et al., 2008).

Variation source	Cross-pollination (%)		
Treatment	2.67	P=0.0224	
CV %	77.73		
Means of treatments	1.12	(± 0.20)	
Covered area with honeybee colony	1.5726	a*	(± 0.35)
Covered area without honeybee colony	0.2016	b	(± 0.05)
Uncovered area for free insect visitation	1.5712	a	(± 0.22)

Means followed by different small letters differ between them by Tukey's test (P<0.05)

Table 2. F values with their respective probability (P), coefficient of variation (CV%) of percentage of resistant plants to glyphosate from seeds of soybean *Glycine max* L. Merrill, variety BRS 133 (conventional soybean) to treatments covered area with honeybee colony, covered area without honeybee colony, and uncovered area for free insect visitation in an experimental field of EMBRAPA Soja, in Londrina-PR city.

Figures 5 and 6 are the flowers of conventional (BRS 133) and transgenic (BRS 245 RR) soybean plants.

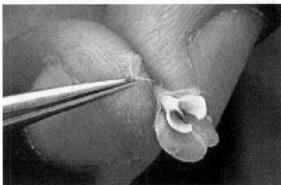

Fig. 5. Flowers of conventional soybean plants – var. BRS 133.

Fig. 6. Flower of transgenic soyben plants var. BRS 245 RR.

The cross-pollination rate was 0.04% in Wisconsin with different varieties of soybean in adjacent rows, in different places (Woodworth, 1922), 0.70% and 0.18% in Virginia, in successive years (Garber & Odland, 1926), and less than 1% in Iowa and Maryland (Weber & Hanson, 1961). In honeybee colony near transgenic canola crops, it was detected that transgene was incorporated to intestinal bacteria chromosomes of the honeybee (Lean et al., 2000).

Furthermore, Ahrent & Caviness (1994) measured the cross-pollination rate in soybean when the plants grown in adjacent lines. The researchers used the colour of flowers, pubescence colour, and leaves fifth foliate as genetic markers. These authors worked with 12 cultivars, in two periods of planting in subsequent years. The cross-pollination rates in this research, probably, were near of maximum because honeybees and other pollinator insects were presented in experimental area during all flowering period. The means in this research varied from 0.09% to 1.63%, these values are averages from two years in different cultivar that differed significantly in relation to cross-pollination. Thus, there is some cultivars have a potential increase to cross-pollination towards to relate earlier.

The dispersion of pollen in soybean was reported by Abud et al. (2003), and they observed more dissemination of trangenic pollen in neighbour lines (0.44% to 0.45%), far way 0.5 m of central parcel, the frequency was reduced drastically in line two (0.04 to 0.14%) and reached zero in line 13, far way 6.5 m of central parcel. Africanized honeybees and other pollinator, not only disseminate the transgenic pollen, but pollinating and insects can carry some plant diseases and can themselves be infected while foraging at flowers (Kevan et al., 2007).

The observation of harvesting behaviour shows that Africanized honeybee has a tendency to collect pollen and nectar in flowers of neighbour plants, in the same line, suggesting that this behaviour can minimize the cross-pollination performed by them, once that cultivar were planted in intercalated lines. Increasing the pollination rate obtained can be explained by intense visitation of *A. mellifera*, means 97.25% of insect visitor in uncovered area. Other bees and Lepidoptera represented, only 1.65% and 1.33%, respectively of visitor to this area.

Evaluations of percentage of GM plants from seeds harvested of conventional plants, in eight directions, far way 1 to 5 m beginnings from pollen source make clear that cross-fecundation average rate observed until 5 m from pollen source was 0.29%, and that was calculated based on number of plants that survived to glyphosate. The cross-fecundation rate reduced from 0.61% far way 1 m of distance to 0.29% far way 2 m, and to 0.23% far way 3 to 5m. The cross-fecundation means, from two to 5 m distance from pollen source did not differ, but when the distance passed from 1 to 2 m, the reduction of cross-fecundation rate was significant. This rate diminished by exponential form, from pollen source, that indicates an allogamy as zero far way 7.76 m.

One of the concerning with advancement of transgenic soybean crops in several regions of Brazil, including the North region of state of Paraná is the occurrence of gene flow. However, Schuster et al. (2007) reported that for a soybean plantation free of grain GM, the isolation recommended is 8 m. Silva & Maciel (2010) reported 0.25% of gene flow between transgenic and conventional soybean plants in Minas Gerais, Brazil. Therefore, the authors reported a potential of contamination, in non-transgenic material, in planting close to RR transgenic soybean or other types of soybean GM that would be released in Brazil.

A mathematic model developed by Cresswell (2010) was used to evaluate the gene flow in experimental fields of safflower pollinated by honeybee *A. mellifera* and *Bombus* spp. This model estimated that the maximum feasible level of bee-mediated, field-to-field gene flow ranged between 0.05% and 0.005% of seed set (95% upper confidence intervals of 0.23% and 0.023%), depending on the composition of the bee fauna. The recommended strategy for minimizing GM gene flow is the utilization of a conventional safflower variety possessing a high capacity for automatic self-fertilization, allowing the conventional plants to grow in large stands to encourage long foraging bouts by bees.

Evaluations of methods for detection of soybean seeds resistant to glyphosate, and gene flow from transgenic to conventional soybean cultivar, performed in Florestal and Viçosa counties, Minas Gerais, Brazil evidenced that seedlings resistant to glyphosate presented cross-fecundation. The highest percentage of hybridization, 1.27% in Florestal and 0.25% in Viçosa counties occurred far way 0.5 m, between pollen source and pollen receptor (Pereira et al., 2007). These rates were about zero far way 2.26 and 1.16 m, to Florestal and Viçosa, respectively. However, the distance is an important factor and able to isolate transgenic soybean fields from conventional fields, preventing the gene escape.

Mathematical models (Baker & Preston, 2003; Cresswell, 2010), empirics (Weekes et al., 2005; Damgaard & Kjellsson, 2005; Gustafson et al., 2006) and deterministic (Colbach et al., 2001; 2004; 2005; Walklate et al., 2004) that simulate the gene flow in fields of trangenic and conventional plants are found in literature. However, each model has its advantages and disadvantages. Some are limited by scarcity of data of environmental variable or spatial scale, others, like empirics are attractive due to relative simplicity and easy utilization, although the deterministic model offers potentially greater capacity predictive and inference. Recent advances in modelling pollen-mediated gene flow in commercial fields are encouraging, but simulating gene flow in heterogeneous landscapes remains an elusive goal. Moreover, practical, user-friendly decision-support tools are needed to inform and guide farmers in implementing coexistence measures (Beckie & Hall, 2008).

4. Conclusion

The Africanized honeybees provided a considerable increase of gene flow from transgenic to conventional cultivar (1.57%), besides these cultivar of soybean were attractive to the honeybees, and that they may perform the cross-pollination in these tested varieties of soybean.

5. Acknowledgements

To the National Council for Scientific and Technological Development (CNPq), process number 479868/01-8, and Coordination of Improvement Staff (CAPES) for their financial support. To Zootechnician (Animal Scientist) Tiago C.S.O. Arnaut de Toledo and Tais S. Lopes, for helping in experimental essays and collecting data.

6. References

Abud, S.; Souza, P.I.M.; Moreira, C.T.; Andrade, S.R.M.; Ulbrich, A.V.; Vianna, G.R.; Rech, E.L. & Aragão, F.J.L. (2003). Dispersão de pólen em soja transgênica na região. *Pesquisa Agropecuária. Brasileira*, Vol.38, No.10, (October 2003), pp. 1229-1235, ISSN 0100-204X

Ahrent D.K. & Caviness C.E. (1994). Natural cross-pollination of twelve soybean cultivars in Arkansas. Crop Science, Vol.34, No.2, pp. 376-378, ISSN 1835-2707

Alves, E.M.; Toledo, V.A.A.; Oliveira, A.J.B.; Sereia, M.J.; Neves, C.A.; Ruvolo-Takasusuki, M.C.C. (2010). Influência de abelhas africanizadas na concentração de açúcares no nectar de soja (*Glycine max* L. Merrill) var. Codetec 207. *Acta Scientiarum – Animal Sciences*, Vol. 32, No. 2, pp. 189-195, ISSN 1807-8672

Baker, J. & Preston, C. (2003). Predicting the spread of herbicide resistance in Australian canola fields. *Transgenic Research*, Vol.12, (December 2003), pp. 731–737, ISNN 0962-8819

Beckie, H.J. & Hall, L.M. Simple to complex: modeling crop pollen-mediated gene flow. *Plant Science*, Vol.175, (June 2008), pp. 615-628, ISSN 0168-9452

Camacho, J.C.B & Franke, B.L. (2008). Efeito da polinização sobre a produção e qualidade de sementes de *Adesmia latifólia*. *Revista Brasileira de Sementes*, Vol.30, No.2, (February 2008), pp. 81-90, ISSN 0101-3122

Chambó, E.D.; Garcia, R.C.; Oliveira, N.T.E. & Duarte-Júnior, J.B. (2010). Application of insecticide and its impact on the visitation of bees (*Apis mellifera* L.) in sunflower (*Helianthus annuus* L.). *Revista Brasileira de Agroecologia,*Vol.5, No.1, pp. 37-42, ISSN 1980-9735

Chiari, W.C.; Toledo, V.A.A.; Ruvulo-Takasusuki, M.C.C.; Oliveira, A.J.B.; Sakaguti, E.S.; Attencia, V.M.; Costa, F.C. & Mitsui, M.H. (2005a). Pollination of soybean (*Glycine max* L. Merril) by honeybees (*Apis mellifera* L.). *Brazilian Archives of Biology and Technology*, Vol.48, No.1, pp. 31-36, ISSN 1516-8913

Chiari, W.C.; Toledo, V.A.A.; Ruvulo-Takasusuki, M.C.C.; Oliveira, A.J.B.; Sakaguti, E.S.; Attencia, V.M.; Costa, F.C. & Mitsui, M.H. (2005b). Floral biology and behavior of Africanized honeybees *Apis mellifera* in soybean (*Glycine max* L. Merril). *Brazilian Archives of Biology and Technology*, Vol.48, No.3, pp. 367-378, ISSN 1516-8913

Chiari, W.C.; Toledo, V.A.A.; Hoffmann-Campo, C.B.; Ruvulo-Takasusuki, M.C.C.; Arnaut de Toledo, T.C.S.O.; Lopes, T.S. (2008). Polinização por *Apis mellifera* em soja transgênica [*Glycine max* (L.) Merrill] Roundup Ready™ cv. BRS 245 RR e convencional cv. BRS 133. *Acta Scientiarum Agronomy*, Vol. 30, No. 2, pp. 267-271, ISSN 1807-8621

Colbach, N.; Clermont-Dauphin, C.; Meynard, J.M. (2001). GENESYS: a model of the influence of cropping system on gene escape from herbicide tolerant rapeseed crops to rape volunteers. II. Genetic exchanges among volunteer and cropped populations in a small region. *Agriculture, Ecosystems & Environment*, Vol.83, (February 2001), pp. 255–270, ISSN 1516-8913

Colbach, N.; Angevin, N.F.; Meynard, J.-M. & Messe´an, A. (2004). Using the GENESYS model quantifying the effect of cropping systems on gene escape from GM rape varieties to evaluate and design cropping systems. *Oléagineux, Corps Gras, Lipides*, Vol.11, No.1, (Jan-Feb. 2004), pp. 11–20, ISSN 1258-8210.

Colbach, N.; Fargue, A.; Sausse, C. & Angevin, F. (2005). Evaluation and use of a spatialtemporal model of cropping system effects on gene escape from transgenic oilseed rape varieties: example of the GENESYS model applied to three co-existing herbicide tolerance transgenes. *European Journal Agronomy*, Vol.22, No. 1, pp. 417–440, ISSN 1161-0301

Cresswell, J. E. (2010). A mechanistic model of pollinator-mediated gene flow in agricultural safflower. *Basic and Applied Ecology*, Vol.11, No.5, (december 2009), pp. 415-421, ISSN 1439-1791

Damgaard, C. & Kjellsson, G. (2005). Gene flow of oilseed rape (Brassica napus) according to isolation distance and buffer zone. *Agriculture, Ecosystems & Environment*, Vol.108, (January 2005), pp. 291-301, ISSN 0167-8809

Embrapa. (2003). Tecnologia de produção de soja – Paraná. In: *Embrapa Soja*, 2003, Available from http://www.cnpso.embrapa.br/download/recomendacoes_parana.pdf

Erickson E.H. (1975). Effect of honey bees on yield of three soybean cultivars. *Crop Science*, Vol.15, No.1, pp. 84-86, ISSN 1835-2707

Free J.B. (1993). *Insect pollination of crops*, Academic Press, ISBN 0122666518, London, UK

Garber R.J. & Odland T.E. (1926) Natural crossing in soybean. *American Society of Agronomy*, Vol.18, pp. 967-970, ISNN 0002-1962

Gazziero, D.L.P.; Adegas, F.S.; Voll, E.; Fornaroli, D. & Chaves, D.P. (2009a). Controle químico de *Conyza spp.*: II População com presença de plantas com mais de 15 cm de altura, *Proceedings of 5th Congresso Brasileiro de Soja*, pp. 67, ISBN: 9788570330130, Goiânia, Goiás, Brasil, 19 to 22 May,

Gazziero, D.L.P.; Voll, E.; Adegas, F.S.; Fornaroli, D. & Chaves, D.P. (2009b). Controle químico de buva: I População com plantas menores de 10 cm de altura, *Proceedings of 5th Congresso Brasileiro de Soja*, pp. 66, ISBN: 9788570330130, Goiânia, Goiás, Brasil, 19 to 22 May,

Gustafson, D.I.; Brants, I.O.; Horak, M.J.; Remund, K.M.; Rosenbaum, E.W. & Soteres, J.K. (2006). Empirical modeling of genetically modified maize grain production practices to achieve European Union labeling thresholds. *Crop Science*, Vol.46, (Sept-Oct. 2006), pp. 2133–2140, ISSN 1835-2707

Hymowitz, T. (1970). On the domestication of the soybean. *Economic Botany*, Vol.24, No.4, (October-December 1970), pp. 408–421, ISSN *0013-0001*

James, C. (2010). *Global Status of Commercialized Biotech/GM Crops: 2010*. Isaaa Brief No.42, ISBN 978-1-892456-49-4, Ithaca, NY, USA

Kevan, P.G. & Eisikowitch, D. (1990). The effects of insect pollination on canola (*Brassica napus* L. cv. O.A.C. Triton) seed germination. *Euphytica*, Vol.45, pp. 39-41, INSS 0014-2336

Kevan, P.G.; Sutton, J. & Shipp, L. (2007). Pollinators as vectors of biocontrol agents – the B52 story. In: *Biological control – a global perspective*. Vincent, C.; Goettel, M.S.; Lazarovits, G. , pp. 319-327, CAB International, ISBN-13: 978 1 84593 265 7, Oxfordshire - UK

Kuroda, Y.; Kaga, A.; Tomooka, N. & Vaughan, D. (2010). The origin and fate of morphological intermediates between wild and cultivated soybeans in their natural habitats in Japan. *Molecular Ecology*, Vol.19, (June 2010), pp. 2346–2360, ISSN 0962-1083

Lean, G.; Angres, V. & Jury, L. (2000). GM genes can spread to people and animals. *The Independent*, (may 2000), London, UK.

Lu, B.R. (2004). Conserving biodiversity of soybean gene pool in the biotechnology era. *Plant Species Biology*, Vol.19, (August 2004), pp. 115-125, ISSN 0913-557X

Malerbo-Souza, D.T.; Nogueira-Couto, R.H. & Couto, L.A. (2004). Honey bee attractants and pollination in sweet Orange, *Citrus sinensis* (L.) Osbeck, var. Pera-Rio. *Journal of Venomous Animals and Toxins including Tropical Diseases*, Vol.10, No.2, p. 144-153, ISSN 1678-9199

McGregor, S.E. (1976). *Insect pollination of cultivated crop plants*, USDA, (Agriculture Handbook, 496), Washington, USA

Mendonça-Hagler, L.C.S. (2001). Biodiversidade e biossegurança. *Biotecnologia, Ciência & Desenvolvimento*, Vol. 18, pp. 16-22, ISSN 1414-6347

Menegatti, A.L.A. & Barros, A.L.M. (2007). Análise comparativa dos custos de produção entre soja transgênica e convencional: um estudo de caso para o Estado do Mato Grosso do Sul. *Revista de Economia e Sociologia Rural*, Vol.45, No. 1, (Mar. 2007), pp. 163-183, ISSN 0103-2003

Paiva, J.G.; Terada, Y. & Toledo, V.A.A. (2003). Seed production and germination of sunflower (*Helianthus annuus* L.) in three pollination systems. *Acta Scientiarum*, Vol.25, No.2, (October 2003), p. 223-227, ISSN 1679-9275

Pasquet, R.S.; Peltier, A.; Hufford, M.B.; Oudin, E.; Saulnier, J.; Paul, L. N.; Knudsen, J.T.; Herren, H.R. & Gepts, P. (2008). Long-distance pollen flow assessment through evaluation of pollinator foraging range suggests transgene escape distances. *PNAS*, Vol.105, pp. 13456-13461, ISSN 0027-8424

Paula Júnior. T.J de & Venzon, M. (2007). *101 Culturas - Manual de Tecnologias Agrícolas*, Epamig, ISBN 978-85-99764-04-6, Belo Horizonte, Brasil.

Pereira, W.A.; Giúdice, M.P.D.; Carneiro, J.E.D.S.; Dias, D.C.F.D.S. & Borém, A. (2007). Fluxo gênico em soja geneticamente modificada e método para sua detecção. *Pesquisa Agropecuária Brasileira*, Vol. 42, No. 7, (julho 2007), pp. 999-1006, ISSN 0100-204X

Ray, J.D.; Kilen, T.C.; Abel, C.A. & Paris, R.L. (2003). Soybean natural cross-pollination rates under field conditions. *Environmental Biosafety Research*, Vol.2, No. 2, (Apr-Jun 2003), pp. 133-138, ISSN 1635-7922

Rubis D.D. (1970). Breeding insect pollinated crops. *Arkansas Agricultural Extension Services*, Vol.127, pp. 19-24

SAS – Statistic Analysis System. (2004). *User's Guide: Statistics*. SAS Inst; Cary, NC; Version 8 (13rd edition).

Schuster, I.; Serra, E.; Vieira, N.; Santana, H. & Sinhorati, D. (2007). Fluxo gênico em soja na Região Oeste do Paraná. *Pesquisa Agropecuária Brasileira*, Vol.42, No.4, (Abril 2007), pp. 515-520, ISSN 0100-204X

Scottish Crop Research Institute. (1999). Gene flow in agriculture: relevance for transgenic crops conference, *Proceedings of a Symposium Held at the University of Keele*, pp. 13-21, ISBN 190139672X, Staffordshire, UK, April 12-14, 1999.

Silva, E.C. & Maciel, G.M. (2010). Fluxo gênico em soja na região sul de Minas Gerais. *Bioscience Journal*, Vol.26, No.4, (Agosto 2010), pp. 544-549, ISSN 1516-3725

Singh, G.; Kashyapr, K.; Khan M.S. & Sharma, S.K. (2001). Effect of pollination modes on yield and quality of hybrid seeds of sunflower, Helianthus annuus L. *Seed Science and Technology*, Vol.29, No.3, p. 567-574, ISNN 0251-0952

Souza, L. (2006). Liberação da soja transgênica no Brasil, vantagem ou não? In: *ANBio*, 7/03/2006, Available from http://www.anbio.org.br/noticias/lucia.htm

Toledo, V.A.A.; Chambó, E.D.; Halak, A.L.; Faquinello, P.; Parpinelli, R.S.; Ostrowski, K.R.; Casagrande, A.P.B. & Ruvolo-Takasusuki, M.C.C. (2011). Biologia floral e polinização em girassol (*Helianthus annuus* L.) por abelhas africanizadas. *Revista Scientia Agrarias Paranaensis*, Vol.10, No.1, (Jan. 2011), pp. 5-17, ISSN 1983-1471

Xu, D.H.; Abe, J.; Gai, J.Y. & Shimamoto, Y. (2002). Diversity of chloroplast DNA SSRs in wild and cultivated soybeans: evidence for multiple origins of cultivated soybean. *Theoretical and Applied Genetics*, Vol.105, (June 2002), pp. 645–653, ISSN 0040-5752

Walklate, P.J.; Hunt, J.C.R.; Higson, H.L. & Sweet, J.B. (2004). A model of pollen-mediated gene flow for oilseed rape. *Proceedings of the Royal Society B: Biological Sciences*, Vol. 271, (Mar. 2004), pp. 441–449, ISSN 0080-4649

Weber C.R. & Hanson W.D. (1961). Natural hybridization with and without ionizing radiation in soybeans. *Crop Science*, Vol.1, pp. 389-392, (November 1961), ISSN 1835-2707

Weekes, R.; Deppe, C.; Allnutt, T.; Boffey, C.; Morgan, D.; Morgan, S.; Bilton, M.; Daniels, R. & Henry, C. (2005). Crop-to-crop gene flow using farm scale sites of oilseed rape (Brassica napus) in the UK. *Transgenic Research*, Vol.14, (July 2005), pp. 749–759, ISNN 0962-8819

Wolff, L.F.B. (2000) Efeito dos agrotóxicos sobre a apicultura e a polinização de soja, citros e macieira, *Proceedings of 13th Congresso Brasileiro de Apicultura, CBA, Confederação Brasileira de Apicultura*, Florianópolis, SC, Brasil, 2000

Woodworth C.M. (1922). The extent of natural cross-pollination in soybeans *American Society of Agronomy*, Vol.14, pp. 278-283, ISNN 0002-1962

Yoshimura, Y. (2011). Wind tunnel and field assessment of pollen dispersal in Soybean [*Glycine max* (L.) Merr.]. *Journal of Plant Research* , Vol.124, No.1, (Jan. 2011), pp. 109-114, ISNN 1618-0860

Permissions

The contributors of this book come from diverse backgrounds, making this book a truly international effort. This book will bring forth new frontiers with its revolutionizing research information and detailed analysis of the nascent developments around the world.

We would like to thank Prof. Dr. Mohammed Naguib Abd El-Ghany Hasaneen, for lending his expertise to make the book truly unique. He has played a crucial role in the development of this book. Without his invaluable contribution this book wouldn't have been possible. He has made vital efforts to compile up to date information on the varied aspects of this subject to make this book a valuable addition to the collection of many professionals and students.

This book was conceptualized with the vision of imparting up-to-date information and advanced data in this field. To ensure the same, a matchless editorial board was set up. Every individual on the board went through rigorous rounds of assessment to prove their worth. After which they invested a large part of their time researching and compiling the most relevant data for our readers. Conferences and sessions were held from time to time between the editorial board and the contributing authors to present the data in the most comprehensible form. The editorial team has worked tirelessly to provide valuable and valid information to help people across the globe.

Every chapter published in this book has been scrutinized by our experts. Their significance has been extensively debated. The topics covered herein carry significant findings which will fuel the growth of the discipline. They may even be implemented as practical applications or may be referred to as a beginning point for another development. Chapters in this book were first published by InTech; hereby published with permission under the Creative Commons Attribution License or equivalent.

The editorial board has been involved in producing this book since its inception. They have spent rigorous hours researching and exploring the diverse topics which have resulted in the successful publishing of this book. They have passed on their knowledge of decades through this book. To expedite this challenging task, the publisher supported the team at every step. A small team of assistant editors was also appointed to further simplify the editing procedure and attain best results for the readers.

Our editorial team has been hand-picked from every corner of the world. Their multi-ethnicity adds dynamic inputs to the discussions which result in innovative outcomes. These outcomes are then further discussed with the researchers and contributors who give their valuable feedback and opinion regarding the same. The feedback is then collaborated with the researches and they are edited in a comprehensive manner to aid the understanding of the subject.

Apart from the editorial board, the designing team has also invested a significant amount of their time in understanding the subject and creating the most relevant covers. They scrutinized every image to scout for the most suitable representation of the subject and create an appropriate cover for the book.

The publishing team has been involved in this book since its early stages. They were actively engaged in every process, be it collecting the data, connecting with the contributors or procuring relevant information. The team has been an ardent support to the editorial, designing and production team. Their endless efforts to recruit the best for this project, has resulted in the accomplishment of this book. They are a veteran in the field of academics and their pool of knowledge is as vast as their experience in printing. Their expertise and guidance has proved useful at every step. Their uncompromising quality standards have made this book an exceptional effort. Their encouragement from time to time has been an inspiration for everyone.

The publisher and the editorial board hope that this book will prove to be a valuable piece of knowledge for researchers, students, practitioners and scholars across the globe.

List of Contributors

Istvan Jablonkai
Institute of Biomolecular Chemistry, Chemical Research Center, Hungarian Academy of Sciences, Budapest, Hungary

Elżbieta Sacała, Anna Demczuk and Edward Grzyś
Wrocław University of Environmental and Life Sciences, Poland

E. Valera
Applied Molecular Receptors Group (AMRg), IQAC-CSIC, Barcelona, Spain
CIBER de Bioingeniería, Biomateriales y Nanomedicina (CIBER-BBN), Barcelona, Spain

A Rodríguez
Micro and Nano Technologies Group (MNTg), Departament d'Enginyeria Electrònica, Universitat Politècnica de Catalunya, C/, Barcelona, Spain

Weiping Liu and Mengling Tang
Zhejiang University, China

Kateřina Hamouzová, Josef Soukup and Pavlína Košnarová
Czech University of Life Sciences Prague, Department of Agroecology and Biometerology, Kamýcká, Prague 6-Suchdol, Czech Republic

Jaroslav Salava and Daniela Chodová
Crop Research Institute, Division of Plant Health, Prague 6-Ruzyně, Czech Republic

Teresa Cegielska-Taras
Plant Breeding and Acclimatization Institute, National Research Institute, Division Poznań, Poland

Tomasz Pniewski
Institute of Plant Genetics Polish Academy of Science, Poznań, Poland

David A. Haukos
U.S. Fish and Wildlife Service, Department of Natural Resources Management, Texas Tech University, Lubbock, Texas, USA

Krithika Muthukumaran, Alyson J. Laframboise and Siyaram Pandey
University of Windsor, Canada

Renato Zanella, Martha B. Adaime, Sandra C. Peixoto, Caroline do A. Friggi, Osmar D. Prestes, Sérgio L.O. Machado and Enio Marchesan
Federal University of Santa Maria, Santa Maria - RS, Brazil

Luis A. Avila
Federal University of Pelotas, Pelotas - RS, Brazil

Ednei G. Primel
Federal University of Rio Grande, Rio Grande - RS, Brazil

Wainer César Chiari, Emerson Dechechi Chambó and Vagner de Alencar Arnaut de Toledo
Animal Science Department, Universidade Estadual de Maringá, Maringá, Paraná, Brazil

Maria Claudia Colla Ruvolo-Takasusuki
Cell Biology and Genetics Department, Universidade Estadual de Maringá, Maringá, Paraná, Brazil

Carlos Arrabal Arias and Clara Beatriz Hoffmann-Campo
Empresa de Brasileira de Pesquisa Agropecuária - Soja, Londrina, Paraná, Brazil